187
Advances in Anatomy Embryology and Cell Biology

Editors

F. F. Beck, Melbourne · F. Clascá, Madrid
M. Frotscher, Freiburg · D. E. Haines, Jackson
H.-W. Korf, Frankfurt · E. Marani, Enschede
R. Putz, München · Y. Sano, Kyoto
T. H. Schiebler, Würzburg · K. Zilles, Düsseldorf

Reviews and critical articles covering the entire field of normal anatomy (cytology, histology, cyto- and histochemistry, electron microscopy, macroscopy, experimental morphology and embryology and comparative anatomy) are published in Advances in Anatomy, Embryology and Cell Biology. Papers dealing with anthropology and clinical morphology that aim to encourage cooperation between anatomy and related disciplines will also be accepted. Papers are normally commissioned. Original papers and communications may be submitted and will be considered for publication provided they meet the requirements of a review article and thus fit into the scope of "Advances". English language is preferred.

It is a fundamental condition that submitted manuscripts have not been and will not simultaneously be submitted or published elsewhere. With the acceptance of a manuscript for publication, the publisher acquires full and exclusive copyright for all languages and countries.

Twenty-five copies of each paper are supplied free of charge.

Manuscripts should be addressed to

Prof. Dr. F. **BECK**, Howard Florey Institute, University of Melbourne,
Parkville, 3000 Melbourne, Victoria, Australia
e-mail: fb22@le.ac.uk

Prof. Dr. F. **CLASCÁ**, Department of Anatomy, Histology and Neurobiology,
Universidad Autónoma de Madrid, Ave. Arzobispo Morcillo s/n, 28029 Madrid, Spain
e-mail: francisco.clasca@uam.es

Prof. Dr. M. **FROTSCHER**, Institut für Anatomie und Zellbiologie, Abteilung für Neuroanatomie,
Albert-Ludwigs-Universität Freiburg, Albertstr. 17, 79001 Freiburg, Germany
e-mail: michael.frotscher@anat.uni-freiburg.de

Prof. Dr. D. E. **HAINES**, Ph.D., Department of Anatomy, The University of Mississippi Med. Ctr.,
2500 North State Street, Jackson, MS 39216-4505, USA
e-mail: dhaines@anatomy.umsmed.edu

Prof. Dr. H.-W. **KORF**, Zentrum der Morphologie, Universität Frankfurt,
Theodor-Stern Kai 7, 60595 Frankfurt/Main, Germany
e-mail: korf@em.uni-frankfurt.de

Prof. Dr. E. **MARANI**, Department Biomedical Signal and Systems, University Twente,
P.O. Box 217, 7500 AE Enschede, The Netherlands
e-mail: e.marani@utwente.nl

Prof. Dr. R. **PUTZ**, Anatomische Anstalt der Universität München,
Lehrstuhl Anatomie I, Pettenkoferstr. 11, 80336 München, Germany
e-mail: reinhard.putz@med.uni-muenchen.de

Prof. Dr. Dr. h.c. Y. **SANO**, Department of Anatomy,
Kyoto Prefectural University of Medicine,
Kawaramachi-Hirokoji, 602 Kyoto, Japan

Prof. Dr. Dr. h.c. T.H. **SCHIEBLER**, Anatomisches Institut der Universität,
Koellikerstraße 6, 97070 Würzburg, Germany

Prof. Dr. K. **ZILLES**, Universität Düsseldorf, Medizinische Einrichtungen,
C. u. O. Vogt-Institut, Postfach 101007, 40001 Düsseldorf, Germany
e-mail: zilles@hirn.uni-duesseldorf.de

A. Nuñez and E. Malmierca

Corticofugal Modulation of Sensory Information

With 16 Figures

Angel Nuñez, PhD
Eduardo Malmierca, MD
Departamento de Anatomía,
Histología y Neurociencia
Faculdad de Medicina
Universidad Autónoma de Madrid
Arzobispo Morcillo 4
28029 Madrid
Spain

e-mail: angel.nunez@uam.es

ISSN 0301-5556
ISBN-10 3-540-36769-1 Springer Berlin Heidelberg New York
ISBN-13 978-3-540-36769-7 Springer Berlin Heidelberg New York

Springer is a part of Springer Science+Business Media
springer.com
© Springer-Verlag Berlin Heidelberg 2007

The use of general descriptive names, registered names, trademarks, etc. in this publication does not imply, even in the absence of a specific statement, that such names are exempt from the relevant protective laws and regulations and therefore free for general use.
Product liability: The publisher cannot guarantee the accuracy of any information about dosage and application contained in this book. In every individual case the user must check such information by consulting the relevant literature.

Editor: Simon Rallison, Heidelberg
Desk editor: Anne Clauss, Heidelberg
Production editor: Nadja Kroke, Leipzig
Cover design: WMX Design Heidelberg
Typesetting: LE-TEX Jelonek, Schmidt & Vöckler GbR, Leipzig
Printed on acid-free paper   SPIN 11782537   27/3150/YL – 5 4 3 2 1 0

List of Contents

Abstract

Sensory signals reach the cerebral cortex after having made synapses in different relay stations along the sensory pathway. The flow of sensory information in subcortical relay stations is controlled by the action of precise topographic connections from the neocortex. Several lines of research indicate that the massive corticifugal system improves ongoing subcortical sensory processing and reorganizes the receptive fields in visual, auditory and somatosensory systems. In all these sensory systems cortical neurons mediate both the highly focused positive feedback to subcortical neurons with overlapping receptive fields and a widespread inhibition to "non-matching neurons". This cortical feedback, which has been called "egocentric selection", can play a pivotal role in gating the sensory information that reaches the thalamus and cortex. Thus, corticofugal projections may contribute to selective attention since they enhance neuronal responses for attentionally relevant stimuli and by suppressing sensory responses of distractive stimuli. Also, corticofugal projections enhance oscillatory activity in order to synchronize neurons located in the same or in different relay stations in order to improve sensory processing. In conclusion, corticofugal pathways precisely control sensory transmission through out the central nervous system.

Abbreviations

AMPA	α-Amino-3-hydroxy-5-methyl-isoxazole-4-propionic acid
AP5	D(-)-2-amino-5-phosphonovaleric acid
APV	D(-)-2-amino-5-phosphonopentanoic acid
CNQX	6-Cyano-7-nitroquinoxaline-2,3-dione
DSCF	Doppler-shifted constant frequency
EPSP	Excitatory postsynaptic potential
GABA	γ-Aminobutiric acid
IPSP	Inhibitory postsynaptic potential
MT	Middle temporal visual area
NMDA	N-methyl-D-aspartate
SI cortex	Primary somatosensory cortex
SII	Secondary somatosensory cortex

1
Introduction

The processing of sensory information by the central nervous system is difficult to understand by the complex interconnections between subcortical and cortical areas. Connections between the thalamus and cortex are largely reciprocal, with information being processed in both feed-forward (periphery to thalamus to cortex) and feedback (corticothalamic) directions. Sensory signals reach the cerebral cortex after having made synapses in different relay stations along the sensory pathway. Classically, excitatory and inhibitory actions have been thought to modulate sensory responses in the relay stations, thereby contributing to sensory processing in the central nervous system. However, anatomical studies demonstrated the existence of precise corticofugal projections to subcortical relay stations a long time ago. Since the late 1950s, electrophysiological studies of sensory cortex activation and inactivation have also shown that the cortex induces excitation and inhibition of sensory responses in subcortical neurons. Although the majority of studies have concentrated on the ascending or feed-forward system, the feedback connections are actually much more numerous than the thalamocortical axons and may contribute to focusing information processing in the visual, auditory, and somatosensory pathways.

Glutamate has been proposed as the neurotransmitter in corticofugal neurons on the basis of early iontophoretic studies (Krnjevic and Phillis 1963) as well as more recent studies with a variety of other techniques (e.g., Fagg and Foster 1983; Rustioni et al. 1983; Donoghue et al. 1985). Consequently, excitatory actions of corticofugal projections on subcortical neurons result from activation of different glutamatergic receptors, while inhibitory actions may stem from activation of local interneurons in the sensory relay stations, which contain γ-aminobutiric acid (GABA).

Here, we review anatomical and physiological data on the function of the corticofugal system in sensory processing and plasticity. First, we will summarize anatomical data that describe precise projections from sensory cortical areas to subcortical relay stations of the same sensory modality. Since the corticothalamic projection is much stronger than the one it sends to other relay stations, we will consider it in a different section. After that, we will show data from electrophysiological studies that demonstrated that sensory cortex modulates neuronal responses at all levels of the sensory pathway. Special attention will be given to the primary somatosensory (SI) cortex modulation of tactile responses in the dorsal column nuclei, as an example of the corticofugal control of the ascending sensory information. Finally, we will consider the functional consequences of this corticofugal control of sensory transmission along the central nervous system.

2
Anatomical Projections from Sensory Cortical Areas to the Thalamus

In this and in the following sections we summarize anatomical evidence on the corticofugal projection patterns to the thalamus. Morphological studies reveal the existence of a huge projection from sensory cortical areas to all relay stations of the same sensory pathway. This precise corticofugal projection strongly suggests that it plays a central role in sensory processing.

2.1
General Characteristics of Corticothalamic Projections

With the exception of olfaction, sensory information is delivered to cortical neurons through excitatory connections made by thalamic cells known as relay neurons. Although the name "relay neuron" might suggest that these cells simply pass synaptic inputs of sensory activity from the periphery to the cortex, it has become increasingly clear that these neurons are members of a complex circuit that involves ascending, descending, and recurrent sets of neuronal connections.

Guillery and Sherman have proposed that thalamic nuclei can be divided into two types: *first order and higher order* (Guillery and Sherman 2002; Sherman and Guillery 2001, 2002). First-order nuclei represent the first transmission to cortex of a particular type of information from the periphery, and higher-order nuclei serve to transmit information between cortical areas via a cortico-thalamo-cortical route. Examples of the former are the lateral geniculate nucleus for vision (relaying retinal input) and the ventral posterior nucleus for somesthesis (relaying medial lemniscal input); examples of the latter are the pulvinar for vision and the posterior medial nucleus for somesthesis. Differences in the corticofugal projections are observed in the first-order and higher-order nuclei (see Sect. 2.2).

Also, to understand the sensory physiology of the thalamus it is important to distinguish the synaptic inputs of thalamic cells. Guillery and Sherman have divided inputs to thalamic cells into *drivers*, which bring the information to be relayed to the cortex, and *modulators*, which serve to modulate thalamic transmission of the driver input (Sherman and Guillery 1998, 2001). Examples of the former are the retinal and medial lemniscal input to the lateral geniculate nucleus and ventral posterior nucleus, respectively. Examples of the latter are brain stem cholinergic inputs from the parabrachial region and feedback projections from layer VI of cortex. All thalamic relays receive a modulatory input from layer VI of cortex, but only higher-order relays receive, in addition, a driver input from layer V. The layer VI modulatory input is mainly feedback, whereas the layer V driver input is feed-forward (Van Horn and Sherman 2004).

The major source of descending input to thalamic relay neurons comes from neurons with cell bodies located in layer VI of the cerebral cortex (Fig. 1). Also, layer V cortical neurons project to the thalamus. These corticothalamic neurons exert both an excitatory and an inhibitory influence on relay neurons, and it is the

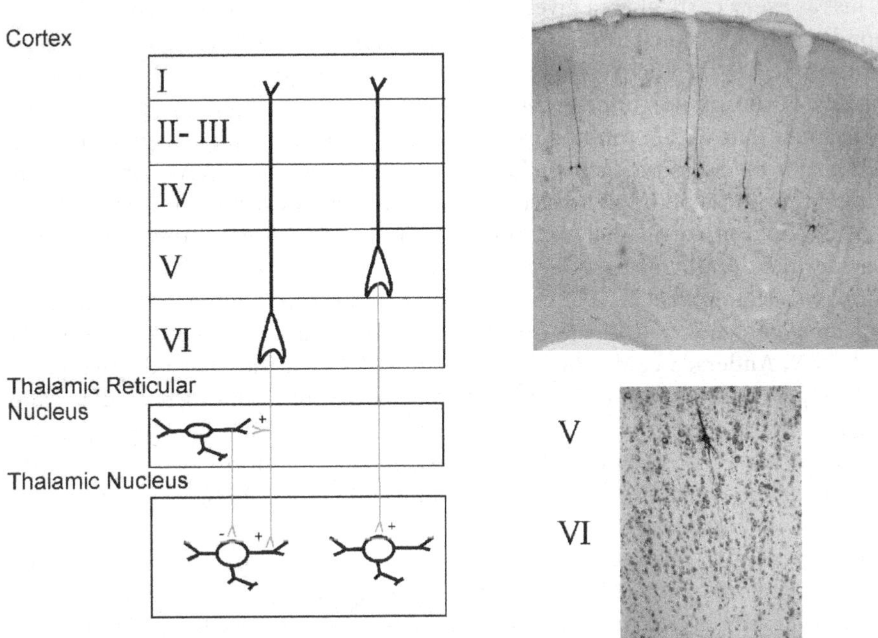

Fig. 1 Schematic representation of corticothalamic anatomical connections. The diagram (*left*) shows layer V and layer VI pyramidal cells projecting to the thalamus. A representative cortical neurons retrogradely labeled from the ventral posterior medial thalamic nucleus is also shown on the *right top*. (Courtesy of C. Avendaño and Pilar Negredo). *Right bottom* shows a pyramidal neuron located in layer V of cortical area 18 projecting to the lateralis medialis thalamic nucleus. (Courtesy of M. Rodrigo-Angulo)

balance of this excitation and inhibition that is thought to influence many of the activity patterns and sensory response properties of relay neurons.

Both sensory and brainstem afferents contact thalamocortical neurons on their proximal dendrites at less than 100 μm from their soma (Wilson et al. 1984; Erisir et al. 1997). In contrast, cortical afferents contact distal dendrites at more than 100 μm from the soma (Erisir et al. 1997), forming a feedback loop that allows the cortex to control the thalamic output that it will receive itself. The location of the corticofugal afferents in the dendritic tree could suggest that their effects are not important in the control of sensory responses of thalamic neurons. However, the corticothalamic projection has been estimated to outnumber the thalamocortical projection by as much as ten to one (Sherman and Koch 1986), indicating that they may be relevant in the modulation of thalamic activity. Typically, most of the corticothalamic fibers are thin (less than 1 μm in diameter; Jones and Powell 1969; Katz 1987; Murphy and Sillito 1996) and therefore conduct slowly (Singer et al. 1975; Tsumoto et al. 1978; Harvey 1980; Tsumoto and Suda 1980; Ahlsen et al. 1982). Moreover, larger-diameter axons are also present.

Just as every cortical area gives rise to a corticothalamic projection, the corresponding thalamic nucleus receives that projection (Jones 1985; Fig. 1). Besides its impressive size, the corticothalamic projection is topographic in all the sensory systems. For example, in the auditory system, there are frequency-specific cortical terminations within tonotopically appropriate medial geniculate nucleus loci. Each major medial geniculate nucleus subdivision receives massive cortical input from four or more of the 12 areas of the auditory cortex (Winer and Larue 1987).

The return of corticothalamic fibers from a cortical area to the nucleus that provides the dominant thalamic input to that area represents a "principle of reciprocity" (Diamond et al. 1969) and the topographic order in the thalamocortical projection matches that in the corticothalamic projection (Guillery 1966; Jones et al. 1979; Andersen et al. 1980). An injection of a dual tracer at a defined point in the representation in the somatosensory, auditory, or visual cortex will usually result in foci of retrograde cell labeling and anterograde fiber labeling that coincide in the ventral posterior, ventral medial geniculate, or dorsal lateral geniculate nuclei (e.g., Jones et al. 1979; Jones 1985; Winer and Larue 1987; Chmielowska et al. 1989).

However, this correspondence is not exact. In most sensory systems, axons descending from a particular part of the topographic map in the cerebral cortex seem to spread beyond the borders of the zone of thalamic cells providing input to that part of the cortical representation. In the ventral posterior nucleus of the mouse, fibers descending from a single barrel representing one whisker in the somatosensory cortex terminate in the barreloid representing that whisker, as well as in the barreloids representing whiskers in adjacent rows on the face (Hoogland et al. 1987).

In the visual system, Murphy and Sillito (1996) examined the distribution of individual corticothalamic axons arising from layer VI cells in the visual cortex and terminating mainly in the A laminae of the dorsal lateral geniculate nucleus of the cat. They discovered that each axon gave rise to 1,300–4,000 boutons, most of which were concentrated in a central zone some 500 μm wide, with a surrounding zone of less dense terminals extending as much as 1,500 μm beyond the central zone. This is much more extensive than the terminal distribution of a retinogeniculate axon (Sur and Sherman 1982; Bowling and Michael 1984). Multiple corticothalamic axons arising from a given retinotopic locations in area 17 converged on an aggregate terminal zone, consisting of a dense central core and thinner surrounding area, with dimensions similar to those of a single fiber. The central core retinotopically corresponded to the cortical locus and matched the maximum region over which the corticogeniculate cells at the corresponding retinal eccentricity could group visual stimuli (approximately 2°; Murphy and Sillito 1987; Grieve and Sillito 1995a, b). However, the zone surrounding the terminals was two to five times larger than the geniculate representation of the area of the visual field providing input to the cortical locus and the maximum region over which the receptive fields of the corticogeniculate cells could group stimuli. Consequently, the corticothalamic axons that arise from cells in this region of the cortex and converge on the dorsal lateral geniculate cells will have different receptive field locations, different orientation, and directional selectivity. This organizational plan is also applied in the corticofugal projection to subthalamic relay stations, as will be described in Sect. 3.

Also, a break in the rule of reciprocity is observed in the existence of bilateral corticothalamic projections, i.e., projections arising in one cerebral cortex and terminating in the ipsi- and contralateral thalamus (Payne and Berman 1984; Molinari et al. 1985). Numerically, the bilateral projection is small compared with the ipsilateral projection and only from specific cortical areas. It appears to arise primarily from cortical areas near the midline and terminates close to the midline of the thalamus, the fibers spilling over, as it were, from terminal fields in the ipsilateral intralaminar, mediodorsal, and ventral medial nuclei. Although small and restricted in its distribution, this bilateral distribution nevertheless seems to have the potential to exert an influence in synchronous activities of the two thalami and cerebral hemispheres.

Although this lack of accurate matching between the cortical and thalamic projections may suggest the existence of nonspecific feedback information, it is more likely that it represents a way to provide thalamic cells with information about events occurring in the surrounding receptive fields and allows sensory response modulation according to the pattern of sensory stimulation. In conclusion, corticofugal projections provide thalamic neurons with sensory information from cortical areas with the same receptive field and from surrounding areas, thereby representing a context over which stimuli can be superimposed. This pattern of organization provides the neuron with ample sensory information to elaborate a synaptic response according to the stimulus characteristics and to the circumstance in which the stimulus occurs.

2.2
Types of Corticothalamic Projections

Based on the intrathalamic distribution of the axonal fields and of terminal morphology, two types of corticofugal fibers arise from the visual, somatosensory, auditory, and motor cortices in mice, rats, cats, and monkeys: fibers arising in either layer V or layer VI cortical cells (Hoogland et al. 1991; Ojima 1994; Bourassa and Deschênes 1995; Bourassa et al. 1995). The layer V corticothalamic projection mainly contacts association thalamic nuclei. In contrast, layer VI corticothalamic projection contacts sensory-specific thalamic nuclei.

The first type of fibers, arising from layer V cortical cells, is a collateral projection issued from long-range axons that project to the brainstem and/or the spinal cord. These axons do not supply a branch to either the thalamic reticular nucleus or the sensory-specific thalamus, although they do approach intralaminar and association nuclei where they form small clusters of large terminals (Royce 1982; Deschênes et al. 1994; Levesque and Parent 1998).

The corticothalamic fibers arising from layer V cells are described as not giving off collaterals in the thalamic reticular nucleus (Harvey 1980; Hoogland et al.

1987; Rouiller and Welker 1991; Ojima 1994; Bourassa et al. 1995). They usually have *en passant* boutons and terminate in grape-like clusters of large (>5 μm) boutons (Hoogland et al. 1987; Rouiller and Welker 1991; Ojima 1994; Bourassa et al. 1995; Rockland 1996). These are thought to form large excitatory synapses virtually identical to those formed by retinal, lemniscal, and other ascending afferent fibers. They end proximally on the thalamic relay cells and are surrounded by the presynaptic dendritic terminals of interneurons, upon which they also terminate, in complex synaptic arrangements known as glomeruli (Hoogland et al. 1987, 1991; Schwartz et al. 1991; Rouiller and Welker 1991; Ojima 1994; Sherman and Guillery 1996; Feig and Harting 1998; Li et al. 2003). Although it is unlikely that the layer V-originating corticothalamic input represents the principal afferent drive to the nuclei that receive it, the location and character of these synapses suggests a more powerful synaptic effect than that mediated by the more typical, layer VI-originating corticothalamic projection (Crick and Koch 1998; Sherman and Guillery 1998).

In general, the corticothalamic projection from layer V returns to thalamic nuclei that provide diffuse inputs, often at layer I, to the cortex (Catsman-Berrevoets and Kuypers 1978; Jones 1985). The primary visual cortex, for example, in addition to its reciprocal projection from layer VI to the dorsal lateral geniculate nucleus (see below), sends a layer V corticothalamic projection to different parts of the lateral posterior–pulvinar complex (Gilbert and Kelly 1975; Lund et al. 1975; Harvey 1980), which are not known to provide inputs to the primary visual area. These nuclei also receive a typical and reciprocal layer VI-originating projection from the cortical areas to which they send their main thalamocortical projection.

In the auditory system, cells in layer V of the primary auditory cortex whose reciprocal projection returns to the ventral medial geniculate nucleus provide a nonreciprocal corticothalamic projection to the dorsal medial geniculate nuclei (Kelly and Wong 1981; Ojima 1994). In the somatosensory nuclei, layer V corticothalamic neurons may provide the projection to the posterior nucleus (Hoogland et al. 1987; Bourassa et al. 1995). Some thalamic nuclei receive only the layer VI-originating projection; the dorsal lateral geniculate, ventral medial geniculate, and ventral posterior nuclei are examples. Other nuclei appear to receive both layer V- and layer VI-originating fibers, e.g., the lateral posterior–pulvinar complex and mediodorsal nucleus.

Thus, data suggest that this type of corticothalamic projection may be involved in a diffuse modulation of thalamic neuronal activity rather than a precise control that could enhance or inhibit specific sensory responses. It is interesting that the nonreciprocating projection, arising from relatively large cells in layer V of the cerebral cortex, is composed of larger and therefore more rapidly conducting fibers than its reciprocal projection. This has suggested that the nonreciprocal corticothalamic projection may serve some alerting function, perhaps directing attention toward relevant stimuli.

The second type of corticofugal fibers, the most numerous, arise from layer VI cells. This type of corticothalamic projection supply axon collaterals to the thalamic reticular nucleus and distribute branches bearing arrays of small terminations across most of the thalamic nuclei. These corticothalamic terminals are presynaptic to the distal dendrites of thalamic relay cells (Jones and Powell 1969; Somogyi et al. 1978; Liu et al. 1995).

The corticothalamic fibers arising from layer VI cells of mouse, rat and cat are thin, characterized by many small (1–2 μm) *en passant* boutons and similarly sized boutons on short stalks, and these fibers form a network in the neuropil (Guillery 1966; Hoogland et al. 1987; Rouiller and Welker 1991; Ojima 1994; Bourassa et al. 1995; Murphy and Sillito 1996; Barlett et al. 2000; Guillery et al. 2001). The number of synapses formed is high and must probably represent the majority of synapses for the dorsal thalamic nucleus, possibly as high as 50% of the total (Wilson et al. 1984; Montero 1991). In fact, corticothalamic synapses account for more than 50% of the total synapses on individual relay cells in the dorsal lateral geniculate (Wilson et al. 1984) and ventral posterior (Liu et al. 1995) nuclei. Synapses are concentrated on distal dendrites, especially of relay cells, where they terminate more frequently than on intrinsic interneurons (Jones and Powell 1969; Montero 1991). Moreover, they give off short collaterals that are distributed in topographic order and end in asymmetrical synapses on the dendrites of thalamic reticular cells (Ohara and Lieberman 1981; Williamson et al. 1993). Little quantification has been done, but qualitative inspection suggests that the density of corticothalamic terminations on thalamic reticular cells may be substantially less than on relay cells in the dorsal thalamus.

The upper part of layer VI contains cells that project exclusively to the sensory-specific nuclei (the dorsal lateral geniculate nucleus or the ventral posterior nucleus), where they form barrel-like or rod-like terminal fields. Those located deeper in layer VI generally exhibit a multinuclear innervation pattern. They innervate large sectors of the associative and/or intralaminar thalamic territories that are affiliated with each of these sensory modalities, the lateral dorsal–lateral posterior nuclei or the posterior group, and participate in the formation of rods or barreloids in specific nuclei (Deschênes et al. 1998). Similar lamina-dependent differences in the distribution of corticothalamic projections have been reported in the auditory system of the cat and in the visual system of the tree shrew (Ojima 1994; Usrey and Fitzpatrick 1996).

2.3
Cells of Origin for Corticothalamic Projections

Most areas of the cerebral cortex, perhaps all, give rise to a corticothalamic projection that arises from larger pyramidal cells located in layer V and layer VI. The organization of layer V projection may be more complex than that of the layer VI-originating projection in the sense that it may not be found in all thalamic relay nuclei and it might form the only corticothalamic projection to other non-relay nuclei; layer V is also the origin of other corticofugal projections to subcortical sites in the midbrain, lower brainstem, or spinal cord of sensory pathways.

Layer V corticothalamic cells are pyramids with medium- to large-diameter triangular somata, a typical basal dendritic spray, and a stout apical dendrite ascending to layer I and giving off oblique branches in most supervening layers (Fig. 2). In comparison with layer VI cells, they have a high spine density. The axon is also typical: the main axon projects to the thalamus via collateral branches and continues on to other subcortical sites (Kelly and Wong 1981; Giuffrida et al. 1983; Ojima 1994; Bourassa et al. 1995). The larger diameter of layer V axons indicates that they conduct faster than do layer VI axons (Harvey 1980; Swadlow and Weyand 1987). Unlike the layer VI corticothalamic cells, however, the axon may not send collaterals to the thalamic reticular nucleus (Hoogland et al. 1987; Rouiller and Welker 1991; Ojima 1994; Bourassa et al. 1995). Within the cortex, the axons of layer V corticothalamic cells give off an extensive system of long horizontal collaterals, mainly in layer V, with intermittent places of terminations in layer III (Ojima 1994). These observations suggest that the layer V corticothalamic cells form part of a system whose actions are more widely distributed in the cortex than those of the layer VI corticothalamic cells.

Since layer V corticofugal projections give off extensive cortical collaterals and projections to non-relay thalamic nuclei, it is possible that corticofugal projections from layer V correspond to a mechanism for directed attention (Ojima 1994) or a feed-forward system from one cortical area to another via the thalamic relay nucleus of the second area (Hoogland et al. 1987; Rouiller and Welker 1991). Layer V corticothalamic projections may also contribute to synchronizing rhythmic activity in large cortical areas, as is indicated below.

Retrograde tracing studies have shown that the vast majority of corticothalamic fibers arise from pyramidal neurons in layer VI of the cerebral cortex (Gilbert and Kelly 1975; Lund et al. 1975; Jones and Wise 1977; Kelly and Wong 1981; Rustioni et al. 1983; Chmielowska et al. 1989; Yeterian and Pandya 1994). Following injection of a retrogradely transported tracer in a particular thalamic nucleus, as many as 50% of layer VI cells will be retrogradely labeled (Gilbert and Kelly 1975; McCourt et al. 1986; Katz 1987). This is also reflected in the relatively large proportion of cells that can be antidromically activated from the thalamus (Tsumoto and Suda 1980). Most of the corticothalamic cells are found in the upper (Jones and Wise 1977) or middle (McCourt et al. 1986) levels of layer VI.

There are a number of morphological features that distinguish layer VI corticothalamic cells from other cortical efferent neurons in general and from layer V corticothalamic cells in particular. The layer VI corticothalamic cells have the small, round cell bodies typical of most layer VI cells, with substantial numbers of horizontally disposed, perisomatic dendrites confined to layer VI (Jones 1975). From one of these or from the soma itself, a slender apical dendrite ascends without much branching into the supervening layers, usually ending in a spray of branches in layer IV (Gilbert and Wiesel 1979; Katz 1987), although a fine branch may even continue on as far as layer I (Jones 1975). Layer VI cortical neurons receive a substantial monosynaptic input from the thalamus (Bullier and Henry 1979; White and Hersch 1982). Although the dendrites are covered in dendritic

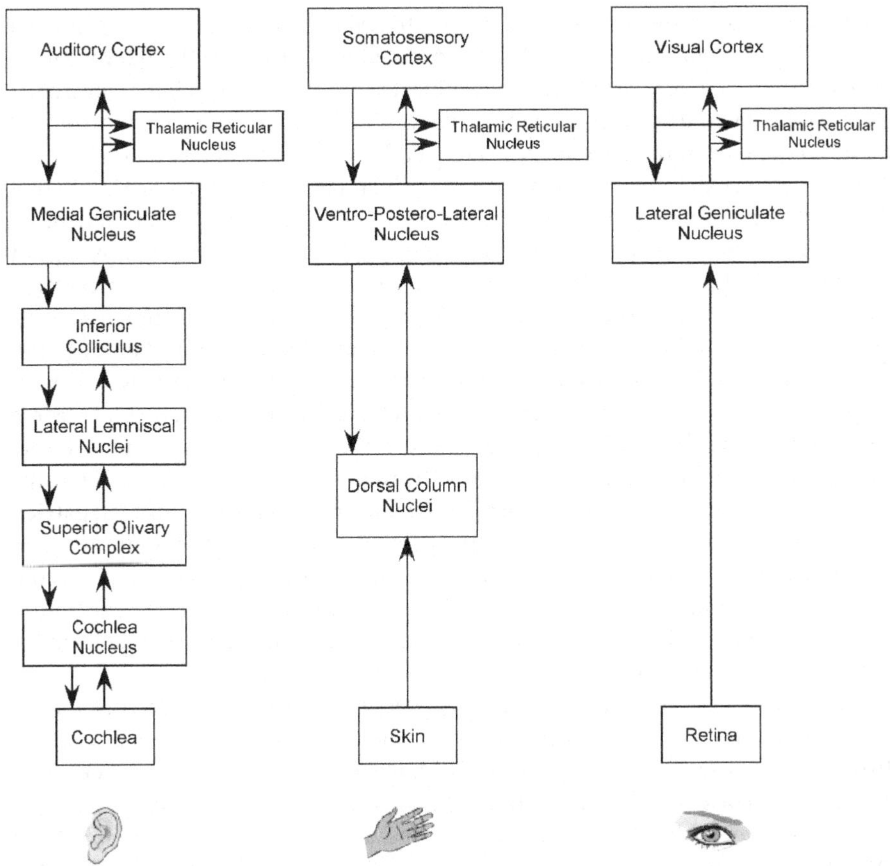

Fig. 2 Schematic representation summarizing the main components of corticofugal pathways in different sensory systems

spines, spine density is relatively low. The axon is perhaps the most typical feature of the layer VI corticothalamic cell. It leaves the cortex and, apart from branches to the thalamic reticular nucleus, terminates only in the appropriate nucleus of the dorsal thalamus. Within the cortex, it gives off relatively few, short, local collaterals in layer VI, but then sends a major recurrent branch to layer IV and the adjacent part of layer III, where it ends in a relatively dense terminal spray (Gilbert and Wiesel 1979; Katz 1987; Ojima et al. 1992; Ojima 1994; Murphy and Sillito 1996). This recurrent branch to layer IV is reportedly not found on other layer VI cells, e.g., those projecting to the claustrum (Katz 1987).

In animals with highly stratified dorsal lateral geniculate nuclei and visual cortices, there may be further specificity in the organization of the corticogeniculate pathway and in the intracortical collateralization of corticothalamic axons. In monkeys, in which corticothalamic cells in a superficial sublayer of layer VI project back to the parvocellular layers of the dorsal lateral geniculate nucleus, the intracortical axon collaterals are distributed particularly to layers IVA and IVCβ, the principal cortical termination layers of thalamocortical fibers arising in the parvocellular geniculate layers. Corticothalamic cells projecting to the magnocellular geniculate layers and located in a deep sublamina of layer VI, by contrast, send intracortical axon collaterals that ramify mainly in layer IVCα, the principal termination site of thalamocortical fibers arising in the magnocellular layers (Lund et al. 1975). In tree shrews, the apparent rule derived from these observations, namely that corticothalamic cells form re-entrant circuits to both the cells of origin and termination of thalamocortical projections can be extended to corticothalamic cells innervating the small-celled population of thalamocortical relay cells. Corticothalamic cells projecting back to the principal layers (I, II, IV, and V) of the dorsal lateral geniculate nucleus are located in the upper part of layer VI and have strong collateral projections to layer IV, where the axons of relay cells in these geniculate layers terminate. Deeper layer VI cells project back to the small-celled geniculate layers (III and VI) and collateralize in layers I–III, the cortical layers where the axons from the cells in the small-celled geniculate layers terminate (Usrey and Fitzpatrick 1996). The terminals of individual corticogeniculate axons and of their stratified collaterals in the visual cortex seem to be concentrated in geniculate layers and sublaminae of layer IV based on ocular dominance, rather than on the segregation of the on-center and off-center channels that are present in the geniculostriate projection in the tree shrew.

Layer VI cortical neurons are not homogeneous. Different types of neurons have been described according to morphological or electrophysiological characteristics. Two kinds of layer VI corticothalamic cell have been morphologically described in the visual cortex of the cat (Katz 1987). Type 1 cells are in the majority, have the typical layer IV bush of dendritic branches, and have axons that are 0.5–1 μm in diameter. Type 2 cells have more restricted dendritic fields that do not reach layer IV, and extremely fine axons, less than 0.3 μm in diameter, with collaterals that are mainly in layer V. These two morphologically distinct types of layer VI corticothalamic cells may correspond, in part, to subpopulations of cells described in terms of their axonal conduction velocities in the cat. Harvey (1980) described two sets of corticothalamic axons: one group, forming about 79% of the population, had slowly conducting axons (the latency to antidromic stimulation from the dorsal lateral geniculate nucleus ranged from 2.6 to 22 ms) that arose from simple cells (according to the receptive field classification of Hubel and Wiesel 1962) and innervated only the dorsal lateral geniculate nucleus. A second group had more rapidly conducting axons (antidromic latency range from 0.52 to 1.3 ms) and arose from complex cells. The presence of slowly and more rapidly conducting corticothalamic axons arising from layer VI cells with simple and complex receptive fields, respectively, has been repeatedly confirmed (Gilbert 1977; Tsumoto and Suda 1980; Harvey 1980b; Ahlsen et al. 1982; Grieve and Sillito 1995a). In all cases, most of the recorded layer VI cells are the slow conducting ones.

Tsumoto and Suda (1980) further divided layer VI corticothalamic cells in the cat visual cortex into three groups on the basis of antidromic latencies from the dorsal lateral geniculate nucleus: 10–40 ms (mean, 20 ms), 3–7 ms (mean, 4.5 ms), and 0.5–2.5 ms (mean, 1.5 ms). When translated into conduction velocities, these gave a slowly conducting group (0.3–1.6 m/s; mean, 0.9 m/s), an intermediate group (3.2–11 m/s; mean, 6.6 m/s) and a fast conducting group (13–32 ms; mean, 20.3 ms). Cells corresponding to all three groups were observed in the binocularly innervated segment of the visual cortex, where they formed 21%, 31%, and 47% of the sample, but the intermediate group was absent from the monocular segment. The population with slowly conducting axons could not be driven by visual stimuli and was located in the deepest part of layer VI. The cells with the fastest conducting axons had receptive fields, which enabled them to be classified as complex cells and were located mainly in the upper part of layer VI. The group with intermediate conduction velocities consisted of simple cells and was mainly located in the middle parts of layer VI of the cortex. These cells probably formed the fastest of the slowly conducting corticothalamic population studied by Harvey (1980) and other authors.

According to electrophysiological properties, Landry and Dykes (1985) described two populations of corticothalamic cells in the SI cortex projecting to the ventral posterior nucleus from layer VI. Approximately 60% of the cells had no spontaneous activity, no obvious receptive field and somewhat slower axons (latencies, 1–18 ms; mean, 5.5 ms). They were located deep in layer VI and bore comparison with the most slowly conducting group observed by Tsumoto and Suda (1980) in the cat visual cortex. The remaining cells were spontaneously active, responded to tactile stimuli applied to definable peripheral receptive fields and had more rapidly conducting axons (latencies, 0.6–7.6 ms; mean, 2.9 ms). They tended to be located more superficially in layer VI. They are comparable to the faster conducting group of corticogeniculate cells described above. Unlike the corticothalamic cells with slowly conducting axons, they could be activated orthodromically by stimulation of the ventral posterior nucleus, but with relatively long latencies.

Also, layer VI projecting cells can be distinguished based on their axonal distributions (Bourassa et al. 1995). Axons arising from cells in the upper part of the layer had anteroposteriorly elongated, rod-like aggregations of terminations in the ventral posterior medial nucleus, possibly conforming to the barreloid organization, although a relationship to the barreloids was not mentioned. The shifted overlap in their terminations does not, however, support the idea of a point-to-point matching of somatotopy, and the findings are thus similar to those in the mouse (Hoogland et al. 1987). Axons arising from cells in the lower part of layer VI terminated in a somewhat more diffuse fashion in the posterior medial nucleus; a few of these also ended in elongated rods in the ventral posterior medial nucleus. All of the layer VI-arising axons were thin and ended in the typical, small, *en passant* and stalked boutons. All gave off collaterals in the somatosensory sector of the thalamic reticular nucleus.

Consequently, all these data suggest the existence of different subpopulations of cortical neurons that project to the thalamus with specific information.

2.4
Neurotransmitter Actions

There is considerable evidence to suggest that driving synaptic inputs and descending corticofugal projections to the thalamus utilize the excitatory amino acid L-glutamate as their neurotransmitter (Bromberg et al. 1981; DeBiasi and Rustioni 1990; DeBiasi et al. 1994; Deschênes and Hu 1990; Broman 1994; Sherman and Guillery 1996; Turner and Salt 1998). Evidence from the lateral geniculate nucleus in cats suggests that the receptors activate directly by retinal inputs are AMPA and NMDA receptors (Scharfman et al. 1990). Medial lemniscal inputs to the ventral posterior nucleus in rats can also activate both AMPA and NMDA receptors (Turner and Salt 1998). The synapses formed by drivers show fairly large excitatory postsynaptic potentials (EPSPs) with paired-pulse depression (Castro-Alamancos 2002; Chen and Regehr 2003; Reichova and Sherman 2004).

The fine fibers of the corticothalamic projection of mouse, rat, and cat enter the dorsal thalamus by traversing the same sector of the thalamic reticular nucleus traversed by the thalamocortical fibers directed toward their cortical area of origin (Jones 1975; Hoogland et al. 1987; Agmon et al. 1995; Bourassa et al. 1995; Murphy and Sillito 1996). Stimulation of corticothalamic inputs to thalamocortical neurons results in monosynaptic EPSPs that involve AMPA and NMDA receptors as well as certain metabotropic glutamate receptors in carnivores and rodents (McCormick and Von Krosigk 1992; Turner and Salt 1998; Castro-Alamancos and Oldford 2002; Li et al. 2003; Reichova and Sherman 2004). The monosynaptic EPSPs on thalamocortical neurons exhibit features that are consistent with the convergence of inputs from many layer-VI cells onto single thalamocortical cells. The ionotropic receptors activated by corticothalamic axons are the same AMPA and NMDA types that are activated by driver inputs. However, because of the very different locations of their synaptic inputs upon the dendritic arbor (driver synapses are found proximally and corticothalamic distally), driver and corticothalamic synapses are unlikely to activate the same individual receptors.

Also, metabotropic glutamate receptors may contribute to corticothalamic synaptic transmission since anatomical studies indicate that metabotropic glutamate receptors are opposed to this synaptic input on distal neurons in sensory thalamic nuclei (McCormick and von Krosigk 1992; Martin et al. 1992; Godwin et al. 1996; Vidnyánszky et al. 1996; Turner and Salt 1998; Reichova and Sherman 2004). Indeed, it has been shown that repetitive stimulation of corticothalamic but not the retinogeniculate input to the dorsal lateral geniculate nucleus of the guinea pig evoked a slow synaptic potential, mediated by the presence of group I, II, and III metabotropic glutamate receptors (McCormick and von Krosigk 1992; Turner and Salt 1999). Thus, corticofugal effects through activation of NMDA and metabotropic receptors may contribute to long-lasting modulatory control of synaptic transmission in the thalamus.

Activation of type I and II metabotropic glutamate receptors on relay thalamic cells increases the production of inositol phosphates and reducing the K^+ leak conductance. This depolarizes the cell, creating an EPSP that is quite slow in onset (>10 ms) and lasts for more than 100 ms (McCormick and von Krosigk 1992; Li et al. 2003; Reichova and Sherman 2004). Also, corticothalamic activation of metabotropic glutamate receptors produces second messenger cascades and release of intracellular Ca^{2+} pools, which may explain the long-lasting effects on relay thalamic cells.

Corticothalamic inputs exhibit a slow EPSP and marked frequency-dependent facilitation that may be important in the integration of inputs (Deschênes and Hu 1990; Alexander et al. 2006). These properties are in marked contrast to ascending inputs, such as those from the retina, which typically neither exhibit frequency-dependent facilitation nor generate slow EPSPs. Accordingly, Li et al. (2003) demonstrated recently in an *in vitro* preparation of rat thalamus that inputs to the lateral posterior nucleus could be divided into two groups based on synaptic properties, one showing synaptic depression and the other synaptic facilitation. They suggest that these inputs were cortical in origin and that one arose from layer V, whereas the other arose from layer VI. Moreover, Reichova and Sherman (2004) demonstrated that inputs from layer VI evoke graded EPSPs with pair-pulse facilitation, and most also showed a type I metabotropic glutamate receptor component in the lateral geniculate nucleus and posteromedial nucleus, while inputs from layer V in the posteromedial nucleus elicited large, all-or-none EPSPs with pair-pulse depression, and without the participation of metabotropic glutamate receptor. Thus, synaptic inputs from cortical layers V and VI may have different influences on thalamic cells. Consequently, results suggest that ascending inputs induce short-lasting EPSPs, mainly through activation of non-NMDA glutamatergic receptors, in order to transmit sensory signals to the cortex. Moreover, corticothalamic projection may be involved in the modulation of these ascending inputs by means of activation of NMDA and metabotropic receptors in the same cell. In the context of corticothalamic feedback, such integration may be important in proposed mechanisms of coincidence detection (e.g., Sillito and Jones 2002).

In addition, corticothalamic or driver inputs generate disynaptic inhibitory postsynaptic potentials (IPSPs) in thalamic cells that involve intranuclear GABAergic interneurons and GABAergic thalamic reticular cells (Ahlsen et al. 1982; Deschênes and Hu 1990; McCormick and von Krosigk 1992; Eaton and Salt 1996; Sanchez-Vives and McCormick 1997; Turner and Salt 1998; Cox and Sherman 2000). Activation of GABA receptors may sculpture sensory thalamic responses.

2.5
Specific Characteristics of Corticothalamic Projections to the Sensory Thalamus

2.5.1
Auditory System

In the auditory system, each major medial geniculate body subdivision receives massive cortical input from four or more of the 12 areas of the auditory cortex, and some are a target of every area (Winer et al. 2001). Besides their impressive size, these pathways are topographic. Among tonotopic fields, this takes the form of frequency-specific cortical terminations within tonotopically appropriate medial geniculate body loci. The projection of nontonotopic and polymodal cortical fields is, unexpectedly, equally topographic spatially, though the functional axes for this order remain unclear in both the cortex and thalamus (Winer and Prieto 2001).

Each of the three principal medial geniculate body nuclear groups receives corticothalamic input. Tonotopic auditory cortical areas target tonotopic medial geniculate body subdivisions preferentially, nontonotopic cortical areas project largely to nontonotopic thalamic nuclei, and polymodal associative medial geniculate body regions, such as the medial division, receive input from all auditory cortical areas and from nonauditory cortex.

The areas of auditory cortex that provide corticofugal modulation to the medial geniculate body form patch-like patterns on the cortical map. The size of the patches ranges from 600 to 1,900 μm in diameter with an average of 1,130 μm (He 1997). This is larger than the spread of the terminal projections of thalamocortical neurons in the cortex, but roughly the same size as the terminal projections of the reciprocal corticothalamic neurons in the medial geniculate body. A possible reason for the observation of such large functional patches in primary auditory cortex is that they reflect the ample ramifications of the corticothalamic projections (He 1997; He et al. 2002).

Subfields for high and low frequencies in the primary auditory area in the rat cortex project to the ventral and dorsal zones of the ventral division of the medial geniculate body, respectively (Hazama et al. 2004). Moreover, collateral projections to the thalamic reticular nucleus appeared topographic in relation to cortical tonotopy. The rule of topography related to cortical tonotopy is applicable to both the primary and nonprimary cortical areas (Kimura et al. 2005). In macaque and marmoset monkeys, the distribution and terminal morphology of the corticothalamic projection originating from the primary auditory cortex show a dense corticothalamic projection in the ventral and dorsal divisions of the medial geniculate body and, to a lesser extent, in the medial division, the posterior thalamic nucleus and the suprageniculate nucleus (Rouiller and and Duriff 2004; De la Motte et al. 2006).

Since the primary cortical auditory areas receive thalamic afferents primarily from the ventral division of the medial geniculate body and the nonprimary auditory areas receive afferents from the dorsal division of the medial geniculate body (e.g., Kimura et al. 2003), the converging direct projections from the tonotopically comparable primary and nonprimary cortical subfields to the ventral division of the medial geniculate body form closed- and open-loop connections. These two circuits may, therefore, individually accomplish different functions, although they could cooperatively excite thalamocortical cells and facilitate thalamic relay of certain frequency sound relevant to cortical processing (Bartlett and Smith 2002; He et al. 2002; He 2003; Xiong et al. 2004). Whereas the projections from the primary auditory subfields appear to serve tonotopy-related feedback modulation within the lemniscal system, the projections from the nonprimary auditory subfields appear to set the lemniscal system attentive to the sound relevant to ongoing nonlemniscal information processing that includes polysensory as well as auditory integration (Komura et al. 2001; Kimura et al. 2004, 2005).

2.5.2
Visual System

The dorsal lateral geniculate nucleus is the principal thalamic relay for retinal signals on their way to the cerebral cortex. It is now well established that the lateral geniculate nucleus functions more as a gate, which regulates retinal information transmitted to the visual cortex, rather than just as a simple relay station for the ascending retinal flow (Singer 1977; Sillito et al. 1994; Cudeiro and Sillito 1996; Marroco et al. 1996; Cudeiro et al. 2000; Montero 2000; Sherman 2001; Guillery and Sherman 2002; Sherman and Guillery 2002; Steriade 2001). One of the key inputs that control the flow of retinal signals at thalamic level is excitatory feedback from layer VI cells of the primary visual cortex (Ahlsen et al. 1982a; Gilbert and Kelly 1975; Lindström 1982; Robson 1983). The number of corticogeniculate synaptic contacts is estimated to constitute about a half (31%–58%; Guillery 1969; Montero 1991; Erisir et al. 1998; Van Horn et al. 2000) of all boutons on relay neurons in the LGN and thus far outweighs the number of synaptic contacts formed by retinal ganglion cells (7%–20%; Montero 1991; Erisir et al. 1998; Van Horn et al. 2000).

In the visual system, the feedback to the lateral geniculate body is retinotopically organized, and the layer VI visual cortex cells that provide the feedback have functionally selective visual-response properties and low spontaneous activity. Thus, the unique signature of any retinal image will in turn evoke a unique pattern of feedback from the visual cortex to the thalamus. An argument that could be made against precise cortical feedback is that the spread of the arborization of the corticofugal axons in the lateral geniculate body is so widespread that any fine distinctions of this type would be simply diffused and lost. Conversely, the presence of asymmetries in the axonal projections of the feedback cortical cells that link to the visual-response properties of the parent thalamic cells supports the concept that spatial focus can be influenced by the feedback cortical projection (see Murphy et al. 1999).

For cats and primates, retinal afferents comprise about 10% of the input to lateral geniculate body relay cells, while the corticofugal feedback connections to the relay cells represent 30% of the input (Erisir et al. 1997; Van Horn et al. 2000; Sherman and Guillery 2002). In the cat, the cortical projection to the lateral geniculate body nucleus, which primarily originates in areas 17 and 18, is dense. Arborization of an individual corticofugal axon has a central core projection of approximately 180–1,080 µm, with a sparse scattering of long-range axons that spread over 500–2,000 µm (Murphy and Sillito 1996; Murphy et al. 2000). Note that the average spread of the retinal X axonal arborization is 150 µm and of the Y axons is 375 µm (Bowling and Michael 1984). Thus, even within their central core, individual corticofugal axons innervate an area of the lateral geniculate body that extends significantly beyond their own location in the retinotopic space. This means that corticofugal axons can influence inputs that may be located outside their own receptive field. Thus, the lack of precision in the cortical feedback may be an advantage to thalamic sensory processing because thalamic neurons receive ample sensory information from the visual cortex that modulates visual thalamic responses according to the context in which the visual stimulus appears. This advantage could be also applied to other sensory systems.

A proper characterization of the visual-response properties of identified corticofugally projecting cells is very important because it identifies the type of stimuli that will influence the transfer of visual information in the lateral geniculate body nucleus and thus sensory responses of primary visual cortex layer VI neurons. In cats, visual system feedback projections originate from both simple and complex cells (Grieve and Sillito 1995a) with simple cells predominating (Gilbert 1977; Tsumoto et al. 1978; Harvey 1980; Tsumoto and Suda 1980; Grieve and Sillito 1995a). Complex cells are spontaneously active as well as being strongly binocular, directional, broadly orientation-tuned, and capable of responding at high stimulus velocities. Simple cells, on the other hand, have little or no spontaneous activity, are sharply orientation-tuned, and include cells strongly or exclusively dominated by one eye (see also Sect. 4.2). Despite the widespread belief that the corticofugal projection originates from cells with very long fields (generally 8° or more, Gilbert (1977) found that the cells that project back to the lateral geniculate body actually have much shorter receptive fields, whereas cells with longer fields project to the claustrum (Grieve and Sillito 1995a). Other investigators have reported color-coded and/or nonoriented cells in layer VI (Livingstone and Hubel 1984). These different types of cells suggest the possibility of a feedback influence that might be segregated for wavelength processing.

2.5.3
Somatosensory System

In the somatosensory system, corticofugal axons from the SI cortex leave the cortex and traverse the striatum in small bundles, which split off into two main streams, as indicated earlier (Veinante et al. 2000). A dorsal stream consists of the axons of layer VI cells, which head directly toward the dorsal thalamus and distribute arrays of small terminations in the thalamic reticular nucleus and the ventral posterior medial and posterior group nuclei. A ventral stream comprises the axons of layer V cells, which continue their course downward through the pallidum and join the internal capsule. At the exit from the pallidum, some fibers give off branches that enter the thalamus to the posterior group nucleus (Veinante et al. 2000).

The reciprocal connections between SI cortex and the ventral posterior lateral nuclei of the thalamus are very precise (Jones et al. 1979). Cortical projections from representations of the same body region in area 3b and area 1 of primates overlap in the single body surface representation in the ventral posterior lateral nucleus of the thalamus (Mayner and Kaas 1986). The topographical relationships between single cortical barrels and thalamic barreloids in the somatosensory system have also been investigated. Following Phaseolus vulgaris-leukoagglutinin and horseradish peroxidase injections restricted to a single barrel column in the mouse, Hoogland et al. (1987) reported that corticothalamic fibers formed rostrocaudally oriented bands in the ventral posterior thalamic nucleus, distributing terminals across a series of barreloids that receive sensory information from vibrissae. Retrogradely labeled cells, however, were found in a unique aggregate that outlined the shape of a single barreloid. Thus, while the thalamocortical feed-forward pathway implies a one-to-one correspondence between thalamic barreloids and cortical barrels, the corticothalamic feedback pathway distributes widely, suggesting that thalamic neurons receive information from cortical neurons with different receptive fields, as also occurs in other sensory systems. In contrast, using horseradish peroxidase to map connections both anterogradely and retrogradely, Land et al. (1995) concluded that the thalamocortical and corticothalamic connections in the rat vibrissa–barrel system were highly reciprocal and that species differences might be at the origin of the discrepancy (Deschênes et al. 1998).

In the rat, convergence of corticothalamic projections has been documented as to those arising from cortical subfields in the SI and secondary somatosensory areas (SII) that represent the same body parts (Alloway et al. 2003). The forms of axonal terminal fields in the ventral posterior complex are similar in the projections of single cortical neurons in SI and SII (Bourassa et al. 1995; Lévesque et al. 1996). Moreover, the axonal terminal fields of single cortical neurons, being restricted to the dimension of a single barreloid in the ventral posterior nucleus (Bourassa et al. 1995), suggest that the convergence takes place in thalamic cells according to somatic representation.

3
Corticofugal Projection to Subthalamic Relay Stations

A similar reciprocity pattern between the corticofugal projections and subthalamic relay stations of the sensory pathway has been described in the auditory, visual, and somatosensory pathways (Fig. 2). Corticofugal fibers from sensory cortical areas make synapses with all the relay stations of the auditory, visual, and somatosensory pathways. Corticofugal projections have two different origins in the cortex according to whether they go to the thalamus (cortical layer VI) or to subthalamic relay stations (cortical layer V). Thus, both corticofugal projections may have distinct functions in the control of sensory processing throughout the sensory pathway. As we will show in detail below, corticofugal projections make it possible to enhance cortically relevant stimuli while decreasing other sensory stimuli.

3.1
Auditory System

Auditory stimuli reach the primary auditory cortex via relays in the cochlear nucleus, the superior olive, the inferior colliculus, and the medial geniculate body (Huffman and Henson 1990). In addition to this ascending pathway from peripheral receptors to the cortex, there is an important corticofugal pathway from the auditory cortex to subcortical relay stations. The finding of equally massive and equally specific descending projections from the cerebral cortex (Diamond et al. 1969) to the medial geniculate body, inferior colliculus (Kelly and Wong 1981; Ojima 1994; Saldaña et al. 1996; Winer et al. 1998; Winer and Prieto 2001), superior olivary complex (Mulders and Robertson 2000), and cochlear nucleus (Weedman and Ryugo 1996) suggests that neurons of the ascending pathways themselves receive significant descending input. These manifold connections suggest that the hierarchical model must now accommodate influences from many sources. Moreover, physiological studies demonstrate that these descending pathways can affect many aspects of subcortical performance, including filtering sharpness of tuning and response plasticity (see Sect. 4.1).

In rats, most cortical neurons projecting to those nuclei appear to target only one of these structures and few cortical pyramidal cells project to both the cochlear and superior olivary complex nuclei (Doucet et al. 2002, 2003). Superficial regions of layer V project to the opposite auditory cortex, and a small number of these neurons project to the inferior colliculus (Games and Winer 1988). Most of the corticocollicular pathway derives from cells distributed in the middle and deep regions of layer V. Layer V is the source of corticobulbar projections to both the cochlear and superior olivary complex (Doucet et al. 2003).

The laminar origin of the corticocollicular projections complement those of the corticothalamic system: they occupy the central and deep part of layer V, where corticothalamic cells of origin are absent; together, these projections fill most of layers V and VI (Prieto and Winer 1999), and their pyramidal cells are among the largest corticofugal neurons (Winer 2005). Their subvarieties have either single spiking or bursting modes of discharge (Hefti and Smith 2000), patterns whose role in processing information likely differs. In monkeys, the projection from the auditory cortex originates from two classes of layer V pyramidal cells. Cells presenting a tufted apical dendrite in layer I have dense terminal fields in the inferior collicular cortices. Pyramids lacking layer I dendritic tufts target the central nucleus of the inferior colliculus in a less dense but tonotopic manner. The caudal cortex projection originates from smaller layer V pyramids and targets the inferior collicular cortices with dense terminal fields. Descending auditory inputs from the core and caudal areas converge in the dorsal and external cortices of the inferior colliculus (Bajo and Moore 2005). Descending connections to the gerbil inferior colliculus form a segregated system in which multiple descending channels originating from different neuronal subpopulations may modulate specific aspects of ascending auditory information.

Corticofugal fibers were originally described as connecting with neurons located in the dorsal and lateral regions, but not in the central region of the inferior colliculus (Faye-Lund 1985). However, other authors using electron microscopy demonstrated that the entire inferior colliculus is the target of the corticofugal input (Saldaña et al. 1996). According to anatomical and neurochemical studies (Feliciano and Potashner 1995), all ipsilateral corticocollicular endings contain round vesicles and form asymmetric synaptic contacts (Saldaña et al. 1996). This suggests that the auditory corticocollicular synapses are excitatory and presumably glutamatergic, since the neurotransmitter for cortical neurons is glutamate (Feliciano and Potashner 1995). Corticofugal inputs preferentially form synapses on the distal dendritic profiles of collicular neurons (Saldaña et al. 1996). Abundant local ipsilateral intranuclear connections of the inferior colliculus may also permit communication across nuclear subdivisions (Saldaña and Merchan 2005).

Functional studies have shown that neocortical electrical stimulation elicits excitatory, inhibitory, and/or complex responses of different latencies in ipsilateral collicular neurons, and these different responses can be explained by the varied synaptic interactions occurring within the inferior colliculus (Watanabe et al. 1966; Mitani et al. 1983; Sun et al. 1989). Since corticofugal synapses are excitatory, the inhibitory and complex responses would indicate that neocortical endings also form synapses with inhibitory, probably GABAergic, neurons (Mitani et al. 1983). Accordingly, virtually all inferior colliculus neuronal types could be targeted by corticofugal axons (Saldaña et al. 1996).

Corticofugal inputs are topographically and tonotopically arranged toward the inferior colliculus (Herbert et al. 1991). Thus, the corticocollicular projections could modulate the processing of auditory information within the inferior colliculus through the facilitation of intracollicular circuits. Corticofugal inputs to excitatory inferior colliculus neurons would favor or amplify the processing of incoming sounds of a given frequency, and the processing of sounds of a different frequency can be attenuated if the corresponding corticocollicular neurons form synapses on inhibitory collicular neurons.

Less is known about auditory cortical relations with the lower brainstem. The proportion of auditory cortical neurons terminating in the superior olivary complex or in the cochlear nucleus is estimated to be no more than 10% of the projection to the inferior colliculus (Doucet et al. 2003). Auditory cortical axons ending in the superior olivary complex concentrate bilaterally in the ventral nucleus of the trapezoid body (Schofield and Coomes 2004). In the cochlear nucleus, ipsilateral labeling predominates, and boutons target the granule cell domain of the dorsal cochlear nucleus, where small terminals and larger mossy fiber endings are concentrated (Schofield and Coomes 2005). These cortical endings contain round synaptic vesicles and form asymmetric synapses on dendritic profiles, implying a modulatory role, probably with an excitatory action, of auditory responses in the first relay station, the cochlear nucleus (Weedman and Ryugo 1996a, b). Input to granule cells arises from neurons in cortical layer V, a layer that is also the origin of projections to the superior olivary complex in both the primary and secondary auditory cortices, and a few of these cortical cells have been observed to project to more than one brainstem target (Doucet et al. 2003).

Moreover, and in regard to direct corticofugal connections with subcortical relay stations of the auditory pathway, increasing evidence suggests that the modulatory corticofugal control of auditory responses is a continuous chain organized into three main steps: first, the corticocollicular projection, from the auditory cortex to the inferior colliculus; second, the colliculus-olivary projection from inferior colliculus neurons that innervate the superior olivary complex; and, third, the olivocochlear projection in which neurons located in the superior olivary complex innervate the acoustic receptors (Vetter et al. 1993; Saldaña et al. 1996). Thus, two distinct parallel descending pathways, directly from either the sensory cortex or the upper relay station, provide precise control of auditory responses at all levels of the auditory pathway.

3.2
Visual System

The visual projection, which extends from the retina to the primary visual cortex, forms synapses in the lateral geniculate body of the thalamus (see above) and in the superior colliculus.

Cortical projections to the superior colliculus involve all of its laminae. The visual cortex projects in a topical manner to the stratum opticum and griseum superficiale and sparsely to the stratum zonale (Lund 1975; Lent 1982; Harting et al. 1992). In monkeys, occipital cortical areas project to superficial layers of the superior colliculus (Kuypers and Lawrence 1967). Retrograde tracer injections in the superior colliculus labeled layer V pyramidal cells in most areas of the neocortex in all existing species (Kawamura and Konno 1979; Schofield et al. 1987; McHaffie et al. 2001). These cortical cells are located superficial to pyramidal cells with projections and project to the spinal cord and the lateral geniculate body. The primary visual area (V1; 18%), secondary visual area (V2; 14%), and middle temporal visual area (MT; 11%), contributed nearly half of the total of labeled cells in the superior colliculus (Collins et al. 2005). Other visual areas that are early in the processing hierarchy provide another 20% of the cortical projections. Inferior temporal visual areas of the ventral stream provide only minor projections (Collins et al. 2005). The results suggest that cortical inputs to the superior colliculus originate predominantly in the early visual areas rather than multimodal or visuomotor areas.

In primates, the MT area is the caudal superior temporal sulcus. This cortical area codes for parameters of moving visual stimuli and projects to deep and superficial strata of the colliculus (Berson 1988; Maioli et al. 1992). It appears to be mainly involved in directing visual attention to novel stimuli of any modality, multimodal integration, and spatial location of the object of interest, followed by the initiation and termination of saccadic eye movements to bring the image of the object of interest onto the high spatial resolution area of the retina—the fovea centralis or area centralis (Meredith et al. 1991; Schiller and Tehovnik 2003). Thus, the superior colliculus appears to be mainly involved in motion detection rather than sensory processing. Consistent with this, in the cat the superficial (retinorecipient) collicular layers receive their principal direct retinal inputs via the W (involved in head and eye movement) and Y (activated by large targets) visual channels (Berson 1988; Tamamaki et al. 1995; Waleszczyk et al. 1999; Wang et al. 2001) and very little or no input via the X-channel, which participates in high-acuity vision (cf. Hoffmann 1973; Tamamaki et al. 1995; for reviews see Berson 1988; Stein and Meredith 1991). Although the retinorecipient, as well as the deep collicular layers, also receive indirect Y-like input relayed via corticotectal projections from layer V of the ipsilateral visual cortex (Hoffmann 1973), there is no clear functional or morphological evidence indicating that the corticotectal projection from the ipsilateral visual cortex relays X-type information (Berson 1988).

On the other hand, it has been clearly demonstrated that the superior colliculus receives a substantial direct input (for review see Harting et al. 1992), not only from the motion-processing extrastriate cortical areas (e.g., area 18 and areas located around the lateral suprasylvian sulcus), but also from the form/pattern-processing extrastriate areas such as area 21a and areas 20a and 20b in the ventral temporal cortex (for review Burke et al. 1998). Results indicate that area 21 in cats exerts a significant, mainly excitatory, influence on the neuronal activity of cells located in the ipsilateral superior colliculus (Hashemi-Nezhad et al. 2003). In most cases, the effect exerted by area 21a is substantially weaker than that exerted by the striate cortex. However, unlike area 17, area 21a strongly modulates the neuronal activity of cells located not only in the superficial (retinorecipient) layers but also in those located in the stratum griseum. This in turn suggests some influence of area 21a on the integrative and motor functions of the superior colliculus. In a proportion of collicular neurons, the feedback projections from area 21a appear to affect specific receptive field properties such as direction selectivity and length tuning.

3.3
Somatosensory System

Processing of touch begins with peripheral inputs from the skin and continues in ascending somatosensory projections at brainstem dorsal column nuclei, thalamic ventroposterior nucleus, and cortical levels. Descending projections from cortex to the dorsal column nuclei and ventroposterior nuclei supplement these ascending projections, but current understanding of cortical influences on subcortical functions remains incomplete.

The dorsal column nuclei, which include the gracilis and cuneate nuclei, are the first relay station in the dorsal column-medial lemniscus pathways in the somatosensory system, which receives sensory information from the hindlimbs and forelimbs. The dorsal column nuclei project to the somatosensory thalamus through the medial lemniscus. The trigeminal nuclei also receive somatosensory

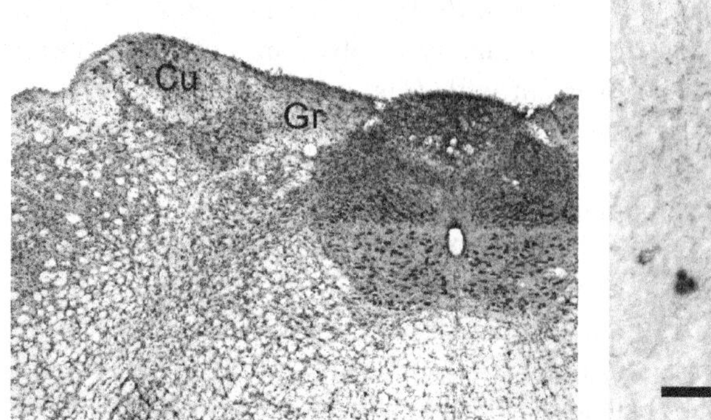
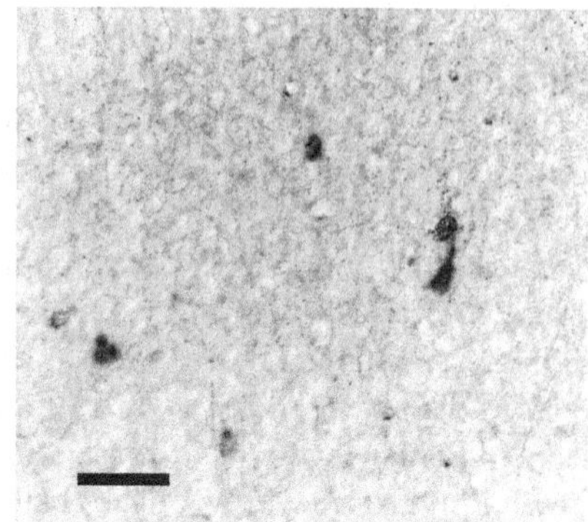

Fig. 3 Microphotograph of a brainstem section showing the gracilis (*Gr*) and cuneatus (*Cu*) nuclei (*left*). Layer V neurons of SI cortex projecting to the dorsal column nuclei (*right*); neurons were retrogradely labeled by cholera toxin injection in the gracilis and cuneatus nuclei. Calibration bar, 60 µm. (Courtesy of M. Rodrigo-Angulo)

information from the face. Dorsal column nuclei neurons, as well as trigeminal neurons, receive two major excitatory inputs that control their function:

1. Ascending somatosensory fibers, which contact both thalamic projection neurons and inhibitory interneurons (Rustioni and Weinberg 1989; DeBiasi et al. 1994; Lue et al. 1996).
2. Corticofugal descending fibers, mainly from cells in the forelimb and hindlimb regions of the SI cortex and to a lesser extent from the second somatosensory area (Jabbur and Towe 1961; Kuypers and Tuerk 1964; Valverde 1966; McComas and Wilson 1968; Weisberg and Rustioni 1976; Rustioni and Hayes 1981; Bentivoglio and Rustioni 1986; Martinez-Lorenzana et al. 2001).

Corticofugal fibers run through the pyramidal tract and their axons branch to the dorsal column nuclei (Martinez et al. 1995; Fig. 3). Recently, magnetic resonance imaging has demonstrated this pathway by anterograde Mn^{2+} transport in the corticospinal and corticothalamic pathways after injection of $MnCl_2$ into the forelimb area of the sensorimotor cortex of rat brains (Allegrini and Wiessner 2003).

Cortical cells projecting to the dorsal column nuclei were confined to the contralateral cortex and their descending axons crossed the midline at the level of pyramidal decussation (Desbois et al. 1999). Cortical projections to the cat dorsal column nuclei arise from layer V pyramidal cells throughout the SI cortex and terminate in the rostral as well as in the cluster regions of those nuclei (Weinberger and Rustioni 1976, 1979; Killackey et al. 1989; Martinez-Lorenzana et al. 2001). Cortical projections from the secondary somatosensory cortex are sparse and confined to the base and rostral region of the cuneate nucleus (Weinberger and Rustioni 1979). Phaseolus vulgaris-leukoagglutinin injections into the SI barrel cortex label pyramidal fibers that terminate at all levels of the contralateral trigeminal complex with the densest terminals being found in laminae III and V of spinal trigeminal nucleus (Jacquin et al. 1990).

The corticofugal system operating on the dorsal column nuclei is powerful and somatotopically arranged (Weinberger and Rustioni 1976, 1979). The ascending dorsal column input terminates mainly on proximal dendrites, whereas the descending corticofugal pathway contacts mainly distal dendrites (Walberg 1966; Rustioni and Sotelo 1974). There is considerable immunohistochemical evidence to suggest that L-glutamate is the excitatory neurotransmitter released by both dorsal column and corticofugal pathways (Rustioni and Cuénod 1982; Banna and Jabbur 1989; Broman 1994; DeBiasi et al. 1994). Moreover, NMDA and non-NMDA receptors have been observed in dorsal column nuclei neurons (Watanabe et al. 1994; Kus et al. 1995; Popratiloff et al. 1997). Additionally, electrophysiological studies of dorsal column nuclei have demonstrated that corticofugal fibers evoke EPSPs with both NMDA and non-NMDA components while the dorsal column input elicits EPSP via activation of non-NMDA receptors (Nuñez and Buño 1999, 2001).

4
Corticofugal Modulations of Sensory Responses

In this section we summarize the evidence that is currently available on the effect of the corticofugal projections on sensory responses of subcortical neurons. We show different results in the auditory, visual, and somatosensory systems that strongly suggest that corticofugal pathways may control ascending sensory transmission. These changes in the activity of midbrain and thalamic neurons have the potential to facilitate the detection of relevant sensory stimuli and may be involved in plasticity changes in sensory systems.

Numerous studies have demonstrated that activation of the cortex induces excitation and/or inhibition of sensory subcortical neurons in auditory, visual, and somatosensory pathways. Initially, physiological data were contradictory. Some authors found predominantly inhibitory activity from the cortex, other authors found excitatory or facilitatory activities, and a third group found roughly equal levels of excitation and inhibition from the cortex to subcortical sensory relay stations. Recently, physiological data demonstrate that this apparent contradiction between previous studies might be resolved if the receptive fields of the stimulated cortical area and the receptive field of the subcortically recorded neurons are considered. Although the mechanisms of corticofugal sensory modulation are similar in the auditory, visual, and somatosensory pathways, we will present these data individually.

4.1
Corticofugal Modulation of Auditory Responses

Numerous studies have concentrated on the neural processing of auditory information along the ascending auditory system. The analysis and representation of sounds in the auditory system have been interpreted as the consequence of divergent and convergent interactions along ascending pathways (Suga 1984; Winer 2005). The presence of large corticofugal projections indicates, however, that the auditory cortex may have an inevitable contribution to auditory information processing in subcortical nuclei (e.g., Diamond et al. 1992; Villa et al. 1991; Ma and Suga 2001; Yan and Ehret 2002). Therefore, many facets of receptive field organization in those nuclei are influenced by the auditory cortex, including threshold, response area, and frequency tuning (Sun et al. 1989; Chowdhury and Suga 2000; Sakai and Suga 2001, 2002).

A long train of repetitive acoustic stimuli comparable to species-specific sounds (Gao and Suga 1998; Yan and Suga 1998; Chowdhury and Suga 2000; Ma and Suga 2001), focal electric stimulation of the primary auditory cortex (Chowdhury and Suga 2000; Ma and Suga 2001, 2003), and auditory fear conditioning (Weinberger 1998; Gao and Suga 2000; Ji et al. 2001) each evoke plastic changes in both the auditory cortex and in subcortical auditory nuclei.

The response of a neuron is usually maximal in magnitude and lowest in threshold at a certain frequency (the best frequency). The change in the best frequency response is called a best frequency shift and it is an example of plasticity in the auditory system. The collicular best frequency shift does not develop when the primary auditory cortex is inactivated during the conditioning (Gao and Suga 1998, 2000), but it is evoked by electric stimulation of the primary auditory cortex (Chowdhury and Suga 2000; Ma and Suga 2001). Therefore, the best frequency shift is modulated by the corticofugal system.

The physiological effects of corticofugal projections to the medial geniculate body of the thalamus and the inferior colliculus of the midbrain have been shown in bats (Sun et al. 1989; Yan and Suga 1996, 1998; Yan and Ehret 2002) and other mammals (Ryugo and Wienberger 1976; Orman and Humphrey 1981; Edeline and Weinberger 1992; Torterolo et al. 1998; Yan and Ehret 2001, 2002). Both facilitation and inhibition of auditory responses have been demonstrated by means of focal stimulation of the auditory cortex (Shun et al. 1989; Torterolo et al. 1998; Zhou and Jen 2000a, b; Jen et al. 2002). Corticofugal facilitation as well as inhibition of neuronal responses to sound stimuli is also observed when the primary auditory cortex is cooled (Ryugo and Weinberger 1976; Orman and Humphrey 1981; Villa et al. 1991).

Electrical stimulation of cortical auditory neurons evokes both facilitation and inhibition of the auditory responses in subcortical neurons. The amount of facilitation and inhibition varies with the frequency of the tone to which the subcortical neurons are responding, and thus the frequency-tuning curves of these neurons can shift their frequency responses (Chowdhury and Suga 2000; Gao and Suga 2000; Ma and Suga 2001). The amount of facilitation and inhibition also varies as a function of the relationship in frequency tuning between the stimulated and the recorded neuron. Corticofugal-evoked facilitation is found in matched collicular neurons, while inhibition predominantly occurs in unmatched neurons (Suga et al. 1997; Torterolo et al. 1998; Yan and Suga 1998; Yan and Ehret 2002). As a result, the response of the recorded neuron is often sharpened in frequency tuning. The collicular processing of sound components in the center of the receptive field (best frequency for this neuron, which induces the largest neuronal response) is largely enhanced, while responses in the surrounding area are suppressed (Yan and Ehret 2002). These frequency-dependent effects improve the input to the stimulated cortical neurons and the subcortical representations of the stimulus parameters to which the cortical neurons are tuned (Chowdhury and Suga 2000; Ma and Suga 2001). This function of the corticofugal system has been named egocentric selection (Yand and Suga 1996; Suga and Ma 2003; see Sect. 5.1).

The effect of electrical stimulation of the auditory cortex on the auditory responses of the medial geniculate neurons has been also studied through in vivo intracellular recordings of anesthetized guinea pigs (Yu et al. 2004). About 50% of the recorded neurons were depolarized and the remaining 50% were hyperpolarized. The corticofugal depolarization of the membrane potential facilitated the auditory responses and spontaneous firing of the medial geniculate neurons. Hyperpolarized neurons showed a decrease in their auditory responses and spontaneous firing. Neurons that were histologically confirmed to be located in the lemniscal medial geniculate nucleus received corticofugal facilitatory modulation, and neurons that were confirmed to be located in the nonlemniscal medial geniculate nucleus received corticofugal inhibitory modulation (Yu et al. 2004). The corticofugal potentiation lasted for an average period of 125 ms (range, 27–400 ms). This period is comparable with the time constant for corticofugal facilitation observed in cat and guinea pig thalami: a few hundred milliseconds (He 1997; He et al. 2002).

Most of the cortical synaptic input into thalamic relay neurons is clearly excitatory and activates AMPA, NMDA, and metabotropic glutamate receptors (Deschênes and Hu 1990; McCormick and von Krosigk 1992; Bartlett and Smith 1999; Tennigkeit et al. 1999). Metabotropic glutamate receptors are coupled to G-proteins and act through the inositol trisphosphate second-messenger pathway, and remain activated for up to several hundred milliseconds (McCormick and von Krosigk 1992; Tennigkeit et al. 1999). The corticothalamic terminals mainly contact distal dendrites. They elicit a slow EPSP with marked frequency-dependent facilitation that may be important in long-lasting sensory integration (Liu et al. 1995a; Bartlett et al. 2000).

Stimulation of the auditory cortex also produced inhibition of thalamic neurons. In a recent extracellular study, He (2003a) observed a mostly inhibitory effect on the ON responses of the nonlemniscal medial geniculate neurons after cortical stimulation. In some cases, the ON response was abolished by cortical stimulation (He 2003a). The average compound IPSP of thalamic neurons produced by cortical stimulation was approximately 11 mV, which was larger than the mean corticothalamic EPSP. The compound IPSP lasted for a long duration of 210 ms. The total effect of the thalamic neurons resulting from cortical stimulation, including the rebounded inhibition, lasted even longer, about 1,000 ms (Yu et al. 2004).

Two potential inputs produce inhibitory potentials in thalamic relay neurons: thalamic inhibitory interneurons or GABAergic thalamic reticular neurons (Houser et al. 1980; Oertel et al. 1983; Yen et al. 1985). Since there are very few interneurons in the thalamus of rodents (<1%; Spreafico et al. 1983, 1994; Winer and Larue 1996; Acelli et al. 1997), it is unlikely that the IPSPs were produced by thalamic interneurons. Thus, most of inhibitory potentials in thalamic relay neurons are elicited by the activity of thalamic reticular neurons in rodents as well as in other mammals.

The major sources of afferent input in the thalamic reticular nucleus are the collaterals of thalamocortical and corticothalamic fibers, and all of them pass through the thalamic reticular nucleus en route to and from the cerebral cortex (e.g., Liu and Jones 1999). The axons arising from the thalamic reticular neurons, after giving off one or two collaterals in the nucleus, only project back to and terminate in the dorsal thalamus but not the cortex (Scheibel and Scheibel 1966; Yen et al. 1985; Steriade et al. 1997). The thalamocortical and corticothalamic collateral terminals display a topographical distribution in the thalamic reticular nucleus that may contribute to inhibiting specific stimuli (Yen et al. 1985).

In the auditory cortex of the mustached bat, sound at roughly 61 kHz is represented in a specialized area called the Doppler-shifted constant frequency (DSCF) area (see Suga 1984; Suga et al. 2002 for reviews). The DSCF area of the primary auditory cortex of the mustached bat is highly specialized for fine frequency analysis. Inactivation of cortical DSCF neurons with lidocaine (a local anesthetic) reduces the auditory responses of thalamic and collicular DSCF neurons that have a matching best frequency to the inactivated neurons and broaden the frequency tuning of these subcortical neurons without shifting their best frequency (Suga et al. 2002; Xiao and Suga 2005). This observation indicates that corticofugal neurons improve auditory signal processing in the frequency domain through their matched subcortical neurons. Such corticofugal modulation is highly specific to matched neurons because it does not occur for unmatched neurons whose best frequency differs by more than 0.2 kHz. On the other hand, inactivation of the cortical DSCF neurons augments the auditory responses of unmatched subcortical neurons at their best frequencies and shifts their best frequency toward the inactivated cortical best frequency. The amount of best frequency shift is proportional to the difference in best frequency between the recorded and inactivated neurons (Zhang and Suga 2000).

Two types of best frequency shift occur in unmatched neurons: centripetal and centrifugal shift. Centripetal best frequency shifts are shifts toward the best frequency of the stimulated cortical neurons, whereas centrifugal best frequency shifts are shifts away from the cortical best frequency (Ma and Suga 2001; Yan and Suga 1998; Chowdhury and Suga 2000; Gao and Suga 2000). Centripetal best frequency shifts increase the number of neurons that respond to a frequency that is the same as the best frequency of the stimulated cortical neurons. That is, they produce an expanded reorganization. In contrast, centrifugal best frequency shifts produce a reduced representation that is associated with the augmentation of responses and the sharpening of tuning curves in matched neurons (Sun et al. 1996). Such reorganization is referred to as a compressed reorganization (Suga et al. 2002). Compressed reorganization increases contrast in the neural representation of an auditory signal; it is therefore presumably more suitable for improving the discrimination of acoustic signals than is expanded reorganization.

Electrical stimulation of cortical neurons evokes centripetal best frequency shifts in the inferior colliculus of big brown and mustached bats (Yan and Suga 1998; Chowdhury and Suga 2000; Gao and Suga 2000; Ma and Suga 2001; Zhang and Suga 2005), mongolian gerbils (Sakai and Suga 2001, 2002) and mice (Yan and Ehret 2002). Shifts in receptive fields of somatosensory cortex neurons are centripetal in different species of mammals, and shifts in orientation selectivity of neurons in the cat visual cortex are also centripetal (see below). Thus, expanded reorganization is common to many mammalian sensory systems.

Since the auditory cortex is largely involved in auditory cognitive processing, it may help to select relevant sounds from the environment. Corticofugal projections may mediate the selective processing of learnt or attended sounds. For example, auditory learning leads to facilitation at a particular locus along the auditory cortical tonotopy that tunes to the frequency of a learned sound (Bakin and Weinberger 1996; Kilgard and Merzenich 1998; Weinberger 1998). Thus, corticofugal modulation may play an important role in learning-induced plasticity in the auditory midbrain as well as the auditory cortex (Gao and Suga 2000; Sakai and Suga 2001).

4.2
Corticofugal Modulation of Visual Responses

The transmission of visual information from the retina to the visual cortex through the lateral geniculate nucleus is a complex process that involves several neuronal mechanisms, elements, and circuits. A prominent feature of the primary visual pathway is the massive feedback connection structure, which arises between the primary visual cortex (area 17) and the lateral geniculate nucleus in the thalamus. Roughly 50% of the synapses on the lateral geniculate cells have their origin in the cortex, whereas direct retinal input constitutes only about 10%–15% (Guillery 1969; Montero 1991; Erisir et al. 1998; Van Horn et al. 2000). To assess the influence of these connections on the sensory processing of visual stimuli is to inactivate the cortex during electrophysiological recordings of lateral geniculate cells. Minor effects were found in most of the older studies that applied this technique and mostly studies reported a rather global depression of lateral geniculate cell activity (Kalil and Chase 1970; Richard et al. 1975; Baker and Malpeli 1977; Geisert et al. 1981). Normally lateral geniculate cells continue to fire as long as the stimulus is presented and little adaptation occurs. This sustained response is substantially reduced during cortical inactivation and the cell response decays to zero rather quickly. A more specific influence in the spatial domain was evidenced by Murphy and Sillito (1987), who showed that the corticofugal connections can have an influence on the length tuning of lateral geniculate cells. Recent experiments have demonstrated that cortical cooling elicits an overall decrease in activity in principal lateral geniculate cells of cats (Waleszczyk et al. 2005). Most cells lowered their spontaneous activity (68% of neurons), and 70% of neurons decrease their peak response amplitudes. In contrast, the 36% of geniculate cells showed an increase in their spontaneous firing and in the peak response to the visual stimulation (24% of neurons). These data taken together suggest that the decreased activity seen in the majority of lateral geniculate cells is due to removal of the direct cortical excitatory input to lateral geniculate cells and/or an increase in inhibition via intrageniculate and/or perigeniculate interneurons. Moreover, Marrocro et al. (1996) demonstrated that the flow of visual information is enhanced because of corticofugal feedback. In particular, they found that the temporal distribution of spikes is altered during cortical cooling. Sillito et al. (1994) showed that lateral geniculate cells fail to synchronize as soon as the cortex is removed. These two findings indicate that the corticofugal connections could predominantly influence time-dependent aspects of cell behavior in the lateral geniculate nucleus.

As the result of detailed investigations by means of extracellular recordings in the dorsal lateral geniculate nucleus, it has become evident that corticothalamic feedback has a number of effects on the transfer functions of this nucleus. Effects on binocular processing in the visual pathway include an influence on feature-dependent, nondominant eye inhibition (Varela and Singer 1987), responses such as surround-mediated attenuation of the center response (Murphy and Sillito 1987) and spatial frequency tuning (Marrocco et al. 1996), enhancements of length tuning (Murphy and Sillito 1987; Sillito et al. 1993), and the temporal structure of geniculocortical spike trains can all be modulated by cortical stimulation (Sillito et al. 1994; Marrocco et al. 1996). Corticogeniculate feedback also influences orientation-dependent modulations of lateral interactions in the nucleus to a considerable extent (Murphy and Sillito 1987; Sillito et al. 1993, 1994; Marrocco et al. 1996), serving to enhance the surround antagonism of a center response in the concentrically arranged receptive fields of geniculate neurons, especially when orientation alignment stimuli are applied to the center and surrounding receptive field (Cudiero and Sillito 1996).

In the feline visual system, feedback projections originate from both simple and complex cortical cells (Grieve and Sillito 1995a) with simple cells predominating (Gilbert 1977; Tsumoto et al. 1978; Harvey 1980; Tsumoto and Suda 1980; Grieve and Sillito 1995a; Priebe and Ferster 2005). The complex cells are spontaneously active as well as strongly binocular, directional, broadly orientation tuned, and capable of responding at high stimulus velocities. The simple cells, on the other hand, have little or no spontaneous activity, are sharply orientation-tuned, and include cells that are strongly or exclusively dominated by one eye. Despite the widespread belief that the corticofugal projection originates in cells with very long fields (generally 8° or more; Gilbert 1977) the cells that project back to the lateral geniculate nucleus actually have much shorter receptive fields, whereas cells with longer fields project to the claustrum (Grieve and Sillito 1995a). Indeed, in layer VI as a whole, short field cells predominate (Grieve and Sillito 1991a, 1995a). One factor stands out from all the available evidence: both the simple and complex cells projecting to the lateral geniculate nucleus tend to be strongly directionally sensitive.

A major characteristic of the influence of feedback on lateral geniculate-cell visual responses seems to be an enhancement of the strength of the inhibitory surround in the presence of moving stimuli, so that cells are more strongly patch-suppressed (and end-stopped) and the excitatory discharge zone for a moving stimulus is more focused (Murphy and Sillito 1987; Sillito et al. 1993; Cudeiro and Sillito 1996; Andolina et al. 2000; Jones et al. 2000a; Webb et al. 2002). Interestingly, the surround suppression seen in the area summation curve for a flashing stimulus is unaffected by the loss of feedback, whereas suppression for the drifting grating is greatly reduced. The enhancement of the inhibitory surround for moving stimuli also seems to lead to increased sensitivity to orientation, direction, and temporal/phase contrasts between the center and surround receptive fields (Sillito et al. 1993; Cudeiro and Sillito 1996; Sillito and Jones 1997; Webb et al. 2002). It is clear that this sensitivity is lost when there is no feedback.

The resulting effects of corticofugal projections on relay thalamic cell discharge can depend upon topographic considerations: stimulation at a visuotopically defined locus in the visual cortex of a cat is followed by both excitatory and inhibitory effects on relay cells situated along retinotopically corresponding projection columns in the dorsal lateral geniculate nucleus (Tsumoto et al. 1978). The inhibitory effect extends into adjacent, nonretinotopically corresponding projection columns, probably because the corticothalamic fibers diverge beyond the corresponding column and because of a similar divergence by axons from inhibitory perigeniculate nucleus cells, which are likely to be the major source of this inhibition (Murphy and Sillito 1996).

Although inhibitory effects mediated by intrinsic inhibitory neurons cannot be ruled out under circumstances in which the corticothalamic pathway is stimulated, it is likely that the perigeniculate nucleus is the principal cause of the inhibition (Waleszczyk et al. 2005). The latency of the inhibition from the cortex is long enough to suggest this; the number of corticothalamic terminals on interneurons appears to be relatively low (Montero 1991), inhibition occurs in species in which intrinsic interneurons are absent or present in only small numbers (Rapisarda et al. 1992; Hu 1993), and, perhaps most importantly, after destruction of thalamic reticular nucleus cells by excitotoxic lesions, corticothalamic stimulation elicits only depolarizing potentials in thalamic relay neurons (Deschênes and Hu 1990).

The very early visual responses seen in the middle temporal visual area (MT; Raiguel et al. 1989; Orban 1997) and its pattern of termination in layers VI-B and VI in the primary visual cortex (Shipp and Zeki 1989; Rockland and Knutson 2000) suggest that it is in a position to exert substantial influence on the early processing of inputs to V1 and the lateral geniculate nucleus. The termination pattern of feedback connections from the MT area in layer VI of the primary visual cortex provides access to these cells associated with feedback for all three processing streams relaying through the lateral geniculate nucleus. Thus, feedback-containing information that is relevant to high-level motion processing in the visual world could influence visual processing for moving stimuli at a very early stage in the development of visual responses in the thalamus. Implanted arrays of three to seven recording electrodes in the lateral geniculate nucleus made it possible to examine the effect of the focal gain changes in the MT area on the responses of the recorded cells. The stimulus used was a drifting texture patch displaced through a sequence of XY coordinates that encompassed the MT field and the lateral geniculate-cell receptive fields. Focal enhancement of the response of the MT area cell produced marked changes in the responsiveness of lateral geniculate cells in the parvo-, magno-, and konicellular layers (Sillito and Jones 2002). These data show that the feedback loop from the middle temporal visual area MT influences the transfer of the retinal input for moving stimuli at the level of the lateral geniculate nucleus and introduces a dynamic of high-level motion processing to the earliest stage in the central visual pathway.

Present data seem to indicate that the visual cortex modulates thalamic activity to promote information transfer through specific locations in the sensory space. The visual cortex is able to increase the gain of its thalamic input within a focal spatial window, selecting key features of the incoming signal. A period of elevated contrast stimulation increases visual responses in lateral geniculate neurons (Cudeiro et al. 2000). These cortical effects are partly due to activation of the type I-metabotropic glutamate receptors, which are only found on distal segments of the relay thalamic cell dendrites, and which enhance the excitatory center of the thalamic receptive field (Cudeiro et al. 2000; Rivadulla et al. 2002).

4.3
Somatosensory System

The flow of sensory information in subcortical relay stations is controlled by the action of precise topographic connections from the neocortex. It is well known that neurons in the somatosensory thalamic nuclei receive dense projections from the SI cortex and that these projections are organized in a topographic manner, as has been described above. However, the function of these projections has been elusive for a long time. Yuan et al. (1985, 1986) reported that inactivation of the SI cortex resulted in reduced thalamic responses to electrocutaneous stimulation without any effect in the thalamus spontaneous activity, stimulus threshold, response latency, and receptive fields. However, other studies reported a facilitatory influence of the SI cortex on evoked thalamic discharges (Andersen et al. 1972; Krupa et al. 1999; Ghazanfar et al. 2001). By mapping the effects of stimulation of the SI cortex, Shin and Chapin (1990) described a range of thalamic effects, including an overall suppressive influence on thalamic sensory responses, which depended upon the topographic location of the neurons in the ventral posterior nucleus of the thalamus. Moreover, Ergenzinger et al. (1998) reported that chronic manipulations of neuronal activity in the primate SI cortex resulted in the expansion of the receptive field in the ventral posterior lateral nucleus. In the same way, elimination of corticofugal projections also led neurons in the spinal subdivision of the trigeminal brainstem complex to increase their responsiveness to whisker stimuli (Jacquin et al. 1990).

Krupa et al. (1999) clearly demonstrated that descending corticothalamic glutamatergic projections influence ascending sensory signals by exerting a dual physiological effect on the thalamic neurons. These effects may include a tonic inhibition, primarily generated by cortically driven excitation of GABAergic neurons located in the thalamic reticular nucleus of the thalamus. Excitation of a topographically related group of thalamic cells and inhibition of adjacent, nontopographically related cells is also seen in the ventral posterior nucleus after focal stimulation of the somatosensory cortex (Rapisarda et al. 1992; Canedo and Aguilar 2000).

The ventral posterior neurons have excitatory receptive fields centered on discrete regions of the body surface. Moreover, indirect evidence suggests that the spatial extent of ventral posterior lateral receptive fields is partly determined by the existence of a suppressive surround area (Ghazanfar et al. 2001). Although the basic properties of thalamic receptive fields are undoubtedly established by feed-forward sensory afferents, some evidence suggests that cortical feedback serves to amplify the effects of sensory stimulation both to the classical receptive field and to the suppressive surround area. Previous studies have shown that the spatial profile of ventral posterior neuronal receptive fields can expand (and sometimes contract) following inactivation of SI cortex (Krupa et al. 1999; Ghazanfar et al. 2001). Thus, at least for some ventral posterior neurons, corticothalamic input appears to operate to sharpen and adjust the profile of thalamic receptive fields, as occurs in the auditory and visual systems.

Involvement of corticofugal pathways in the process of temporary deafferentation-induced thalamic plasticity has been suggested previously. A temporary block of the SI cortex with muscimol (GABAergic agonist) significantly reduced the number of neurons, which showed temporary deafferentation-induced plasticity, in the ventral posterior medial thalamus (Krupa et al. 1999). These authors suggested that the descending dual influences of excitation and inhibition may involve the GABAergic neurons in the thalamic reticular nucleus and the direct glutamatergic projections from the SI cortex. Results imply that temporary deafferentation of a small area of the peripheral receptive field may induce system-wide plasticity involving corticothalamic modulation (Jung and Shin 2002). Therefore, corticothalamic projections are capable of modulating ascending sensory volleys originating in peripheral receptors and they provide a fundamental component that drives the immediate thalamic reorganization observed after a peripheral sensory deafferentation.

The dorsal column nuclei, which include the gracilis and cuneate nuclei, are the first relay station in the somatosensory system. Dorsal column nuclei neurons receive information from peripheral receptors located in hindlimbs and forelimbs and send that information to the somatosensory thalamus. Also, these neurons receive corticofugal-descendant fibers through the pyramidal tract; these fibers arise from layer V pyramidal cells in the SI cortex and, to a lesser extent, from the second somatosensory area, as well as from the motor cortex (Jabbur and Towe 1961; Valverde 1966; Weisberg and Rustioni 1976; Rustioni and Hayes 1981; Cheema et al. 1983; Martinez-Lorenzana et al. 2001). A similar projection to the trigeminal nucleus has been demonstrated as well. Results indicate the existence of a corticofugal projection from the face area of the SI cortex to the spinal trigeminal complex and that the projection is somatotopically organized (Dunn and Tolbert 1982).

Corticofugal projections were first demonstrated by electrical stimulation of the cortex. These original studies showed that excitatory and inhibitory actions can be elicited in dorsal column nuclei cells by electrical stimulation of the somatosensory-motor cortex (Towe and Jabbur 1961; Gordon and Jukes 1964; Lewitt et al. 1964; Cheema et al. 1983; Cole and Gordon 1992; Malmierca and Nuñez 1998, 2004). Peripheral and cortical synaptic inputs evoke EPSPs and IPSPs in neurons recorded *in vivo* (Andersen et al. 1964a, b; Schwartzkroin et al. 1974; Canedo and Aguilar 2000; Mariño et al. 2000) or *in vitro* (Nuñez and Buño 1999, 2001).

There is considerable immunohistochemical evidence to suggest that L-glutamate is the excitatory neurotransmitter released by dorsal column and corticofugal pathways in the dorsal column nuclei (Rustioni and Cuénod 1982; Conti et al. 1989; Banna and Jabbur 1989; Broman 1994; DeBiasi et al. 1994). Moreover, histological studies in the rat have identified NMDA and non-NMDA receptors in the dorsal column nuclei (Watanabe et al. 1994; Kus et al. 1995; Popratiloff et al. 1997). These results have been confirmed by electrophysiological studies *in vitro* (see below).

In the following part, we will show results that demonstrate the effect of corti-cofugal projections from the SI cortex on the dorsal column nuclei cells and the modulation of their tactile responses. Mechanisms of this corticofugal modulation have been studied in *in vivo* and *in vitro* preparations (Malmierca and Nuñez 1998, 2004; Buño and Nuñez 1999, 2001). These results are shown in the following paragraphs.

In urethane anesthetized rats, gracilis neurons display different responses to cortical stimulations according to the cortical site of stimulation: receptive fields of the stimulated cortical area and the gracilis cell overlapped (matching) or receptive fields were nonoverlapping (nonmatching). Electrical stimulation of the SI cortex evoked one to three spikes in most gracilis cells with matching receptive fields (Fig. 4A). In contrast, cortical stimulation evoked smaller responses in gracilis neurons with nonmatching receptive fields (0.7 ± 0.05 and 2.1 ± 0.12 spikes/stimuli, respectively; Fig. 4B). These results agree with those of Cole and Gordon (1992), who reported that the lowest threshold for eliciting spike firing with cortical stimulation was found in dorsal column nuclei neurons with a receptive field that corresponded to the position of the cortical electrode within the hindpaw representation; the threshold increased progressively for cells with more distant receptive fields. These observations are conceivably due to the existence of a well-organized topographic projection between the somatosensory cortex and the dorsal column nuclei (Weisberg and Rustioni 1976; Martinez-Lorenzana et al. 2001). This

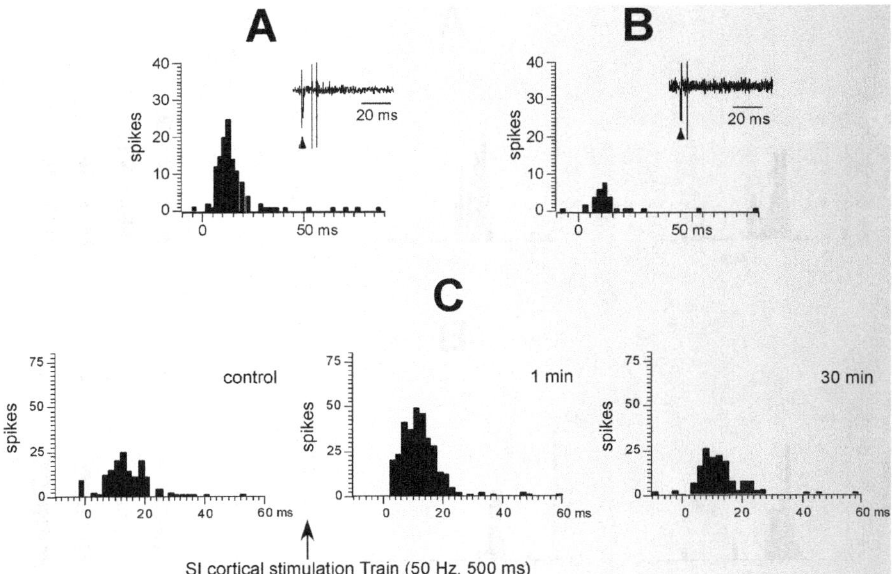

Fig. 4A–C Gracilis cellular responses to primary somatosensory cortical stimulation. **A** Peristimulus time histogram (PSTH) of responses evoked by cortical stimulation in a representative gracilis neuron with overlapping receptive fields of the cortical stimulated area and the gracilis cell. **B** PSTH of the cellular responses evoked by cortical stimulation a typical gracilis neuron with nonoverlapping receptive fields. *Insets* show raw data. *Vertical arrowhead* indicates cortical stimulation. PSHTs are the sum of 40 trials. (Modified from Malmierca and Nuñez 2004)

fine focusing of the corticofugal pathway would be expected to have a functional significance in spatial discrimination and attention processes.

A train of cortical stimuli (0.3 ms at 50 Hz for 500 ms) generated a short-lasting facilitation of its responses when receptive fields overlap. The increment of cortically evoked responses was statistically significant 1 min later and returned to control values 15–30 min after the application of the stimulation train (Fig. 4C). In neurons with nonoverlapping receptive fields, a train of cortical pulses did not significantly affect either the cortical responses or their spontaneous firing rate. Thus, cortical stimulation induced a homosynaptic facilitation of its responses.

Besides of the homosynaptic facilitation, cortical stimulation also induced changes in the tactile responses of dorsal column nuclei (heterosynaptic facilitation). Application of a train of cortical pulses (0.3-ms pulses at 100 Hz for 500 ms) also induced a short-lasting facilitation of tactile responses only in neurons with matching receptive fields (Fig. 5A). The facilitation was observed immediately after the cortical train stimulation (1 min) and at 5 min later in gracilis neurons. It consisted in an increase of the number of spikes evoked by the tactile stimuli (Fig. 5A). Control values had returned 15 min after the cortical stimulating train. In contrast, cortical train stimulation inhibited tactile responses in most gracilis cells

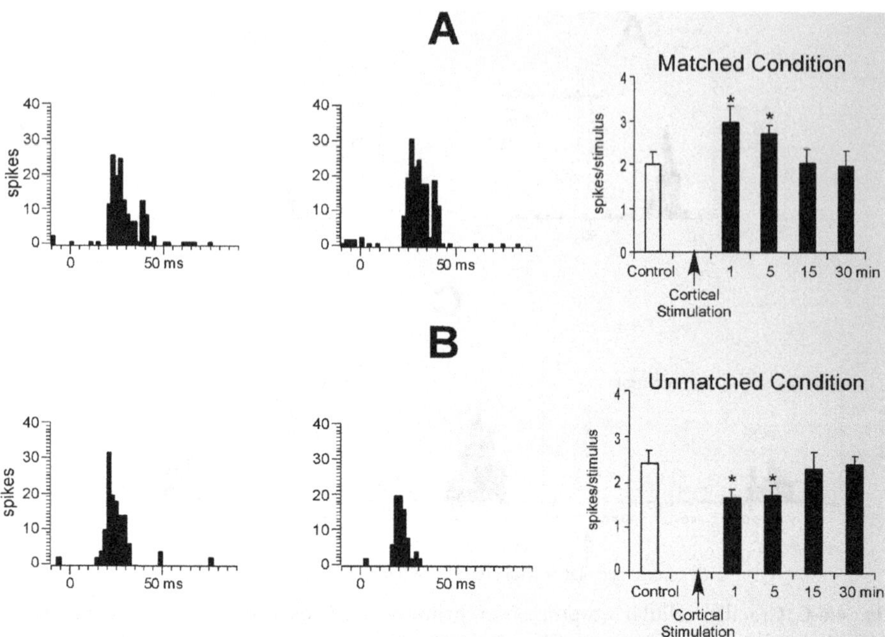

Fig. 5A, B Time course of cortical stimulation effects on gracilis tactile responses. **A** Responses evoked in a gracilis neuron by tactile stimuli. PSTHs show tactile responses before (*left*) and 1 min after (*right*) a cortical stimulus train (0.3 ms at 100 Hz for 500 ms) in cells with an receptive that overlapped the receptive of the stimulated primary somatosensory cortical area. Plot on the *right* shows the mean tactile responses at different time intervals. Note that gracilis neurons were facilitated by cortical stimulation. **B** Is similar to histograms and plot **a** but in cells with nonoverlapping receptive fields. Note that gracilis neurons with nonoverlapping receptive fields were inhibited by cortical stimulation. Plot on the right shows the mean tactile responses. PSHTs are the sum of 40 trials. (* $p<0.05$; ** $p<0.001$), as in the following figures. (Adapted from Malmierca and Nuñez 2004)

with nonmatching receptive fields (Fig. 5B). Consequently, the corticofugal projection may play an important role in the short-term plasticity of the somatosensory system since cortical stimulation induces facilitation or inhibition according to the activated cortical area.

Fig. 6A, B Changes in the receptive field of gracilis neurons evoked by cortical stimulation with overlapping receptive fields. **A** A drawing of the receptive field of a representative gracilis neuron and the PSTHs of the responses evoked by tactile stimuli delivered in two different sites (site 1: functional center of the receptive field; site 2: periphery of the receptive field). **B** One minute after a train of cortical stimuli (0.3 ms at 100 Hz for 500 ms) tactile responses from the functional center increased but they were decreased in the periphery, eliciting a reduction in the size of the receptive field. The *stippled area* in the drawing indicates the RF. PSHTs are the sum of 20 trials. (From Malmierca and Nuñez 2004)

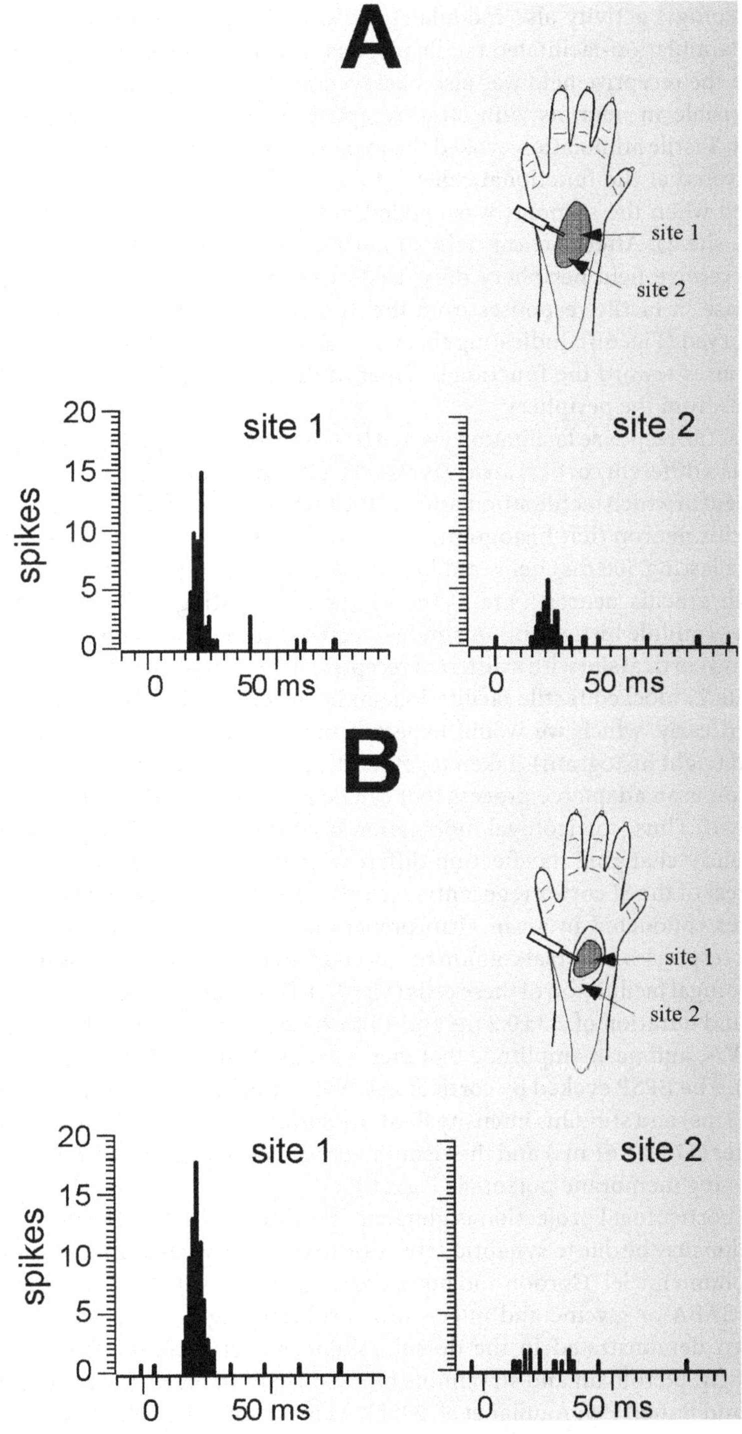

A
site 1
site 2
20
15
10
5
0
spikes
site 1
0
50 ms
site 2
0
50 ms
B
site 1
site 2
20
15
10
5
0
spikes
site 1
0
50 ms
site 2
0
50 ms

Corticofugal activity also modulates the size of the receptive field. In case of cortical stimulation-facilitated tactile responses (matching receptive fields), a decrease in the receptive field was also observed in gracilis neurons. This effect was clearly visible in neurons with large receptive fields on the central pad of the hindfoot. Tactile stimulation evoked the maximal unit response when the stimulus was delivered at the functional center of the receptive field (Fig. 6A, site 1) and decreased when the stimulus was applied to the periphery of the receptive field (Fig. 6A, site 2). After cortical train stimulation, responses to tactile stimulation of the receptive field periphery decreased or even disappeared. Simultaneously, an increase in tactile responses from the functional center of the receptive field was observed (Fig. 6B), indicating that cortical activation tends to focus the cellular responses toward the functional center of the receptive field, reducing tactile responses from the periphery.

The tactile response facilitation evoked by the corticofugal pathway was blocked as soon as a different cortical area was also stimulated. Figure 7 shows an illustrative experiment in which tactile stimulation of the first toe induced a moderate response in a gracilis neuron (left histogram). A train of electrical pulses (0.3 ms duration at 100 Hz lasting 500 ms) delivered in a SI cortical area with a matching receptive field with gracilis neuron (site 1) induced a short-lasting facilitation of tactile responses (middle histogram). A similar electrical stimulation applied 1 min later to a second cortical site with a different receptive field (receptive field in the forepaw Fig. 7; site 2) blocked tactile facilitation, and control tactile response values were recovered early, which we would expect if the second cortical stimuli had not occurred (right histogram). Taken together these results indicate that corticofugal facilitation is an adaptable process that can be reversed if a different cortical area is activated. Thus, corticofugal modulation of dorsal column nuclei cells may be continuously changing its effect on different neuronal populations according to the interest of the SI cortex (egocentric selection; see Malmierca and Nuñez 2004).

Studies conducted in an in vitro preparation revealed the characteristics of synaptic responses in dorsal column nuclei and the cellular mechanisms implicated in corticofugal facilitation of these cells (Fig. 8A). Dorsal column EPSPs had a mean latency and duration of 2.3±0.2 ms and 10.6±2.3 ms, respectively, a rising slope of 3.7±0.5 V/s, and peak amplitude that increased gradually with stimulus intensity (Fig. 8B). The EPSP evoked by corticofugal fiber stimulation had a similar latency (2.9±0.48 ms) and stimulus intensity EPSP amplitude relationship, but the duration was longer (17.5±1.61 ms) and the rising slope was slower (2.7±0.4 V/s) when held at the resting membrane potential (Fig. 8C).

Since corticofugal projection is glutamatergic, the inhibitory effects of cortical stimulation may be due to synaptic activation of inhibitory interneurons within the dorsal column nuclei (Gordon and Jukes 1964; Andersen et al. 1964b). Neurons that contain GABA or glycine and interneurons colocalizing both neurotransmitters have been demonstrated in the dorsal column nuclei (Popratiloff et al. 1996). Accordingly, picrotoxin and strychnine blocked IPSPs in the dorsal column nuclei (Nuñez and Buño 1999; Aguilar et al. 2003). As is shown in Fig. 8D, dorsal column-

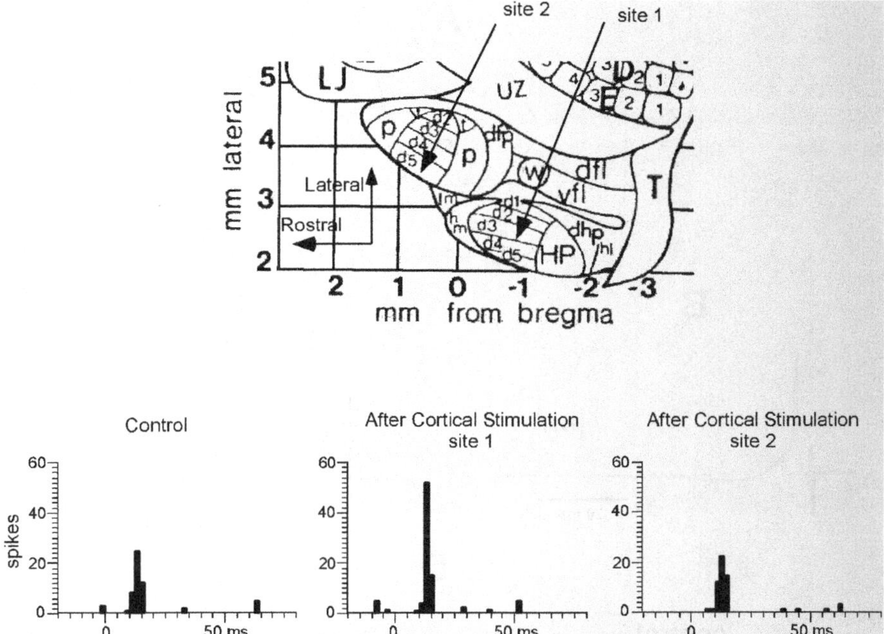

Fig. 7 Effect of stimulation of different primary somatosensory cortical areas on tactile gracilis responses. A schematic drawing of the stimulated cortical areas and PSTHs of tactile responses in a representative gracilis neuron are shown. Stimulation cortical electrode in site 1, but not in site 2, had the same RF as the recorded gracilis neuron. A train of cortical stimuli (0.3 ms at 100 Hz for 500 ms) at site 1 facilitated tactile responses. One minute later, equal cortical train stimulation applied at site 2 blocked the tactile response facilitation. PSHTs are the sum of 40 trials. (Modified from Malmierca and Nuñez 2004)

EPSPs were followed by an IPSP. Bath-applied picrotoxin (a GABA$_A$ antagonist; 30 μM) blocked the IPSP.

The most distinctive difference between dorsal column and corticofugal EPSPs was the voltage dependence of the latter and the activation of different glutamatergic receptors. Stimuli applications at different membrane potentials evoked the expected decrease in dorsal column EPSP amplitude with depolarization without changing the EPSP time course. The dorsal column EPSPs were not affected by APV (50 μM; an NMDA receptor antagonist), but were reversibly blocked by CNQX in all cases (10 μM; a non-NMDA receptor antagonist; Fig. 9A), indicating that non-NMDA receptors were activated by electrical stimulation of the dorsal column and, reasonably, by tactile sensory stimuli that go through the dorsal column fibers.

In contrast, corticofugal-EPSPs displayed a late voltage-dependent component that was absent at hyperpolarized potentials but that appeared at the resting level, as suggested by the longer duration of corticofugal EPSP than dorsal column EPSP (Fig. 9B). Superfusion with APV reduced the corticofugal EPSPs in a voltage-dependent manner. Inhibition by APV was small or even absent at hyperpolarized

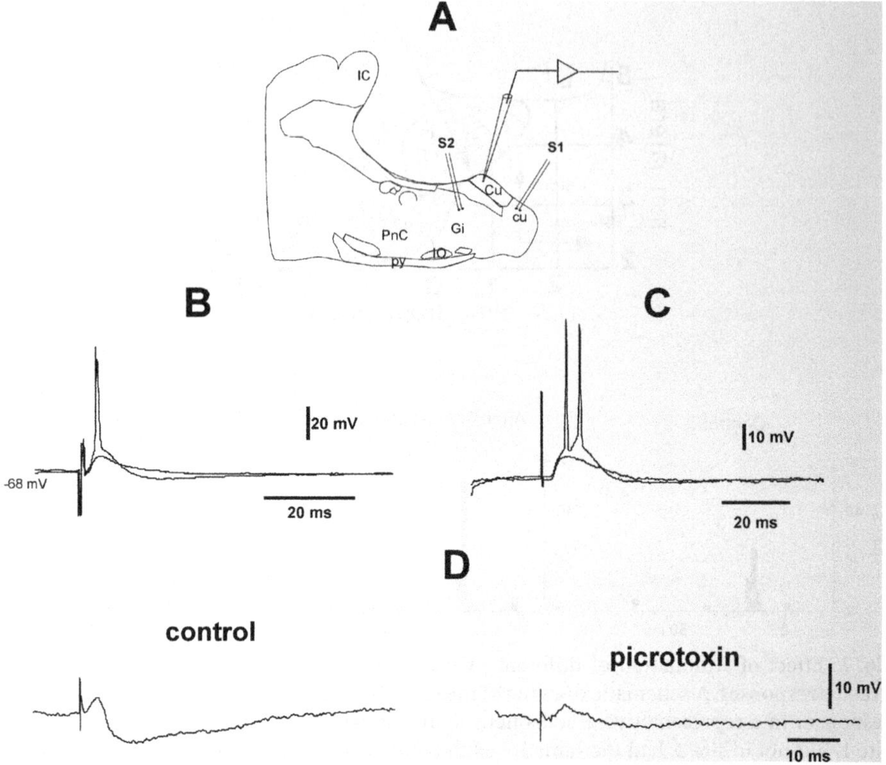

Fig. 8A–D EPSPs evoked by dorsal column and corticofugal fiber stimuli in dorsal column nuclei neurons recorded in vitro. **A** Diagram of recording and stimulating electrode locations. **B** EPSPs evoked by dorsal column stimuli. The EPSP amplitude increased with stimulus intensity and the neuron fired when the threshold was reached. **C** EPSPs evoked by corticofugal fiber stimuli. **D** IPSPs were eliminated by picrotoxin (30 μM). Abbreviations in **A**: *Cu*, cuneate nucleus; *cu*, cuneate fasciculus; *Gi*, gigantocellular reticular nucleus; *IC*, inferior colliculus; *IO*, inferior olive; *PnC*, caudal pontine reticular nucleus; *py*, pyramidal tract; *Sol*, nucleus solitary tract. (Adapted from Nuñez and Buño 1999)

membrane potentials, but the corticofugal EPSPs were markedly reduced at depolarized membrane potentials. Adding CNQX (10 μM) to the solution containing APV abolished the corticofugal EPSPs. Wherefore, dorsal column EPSPs were mediated via activation of non-NMDA receptors but not NMDA receptors, while the corticofugal EPSP resulted from the activation of both non-NMDA and NMDA receptors.

Interactions between excitatory synaptic responses evoked by activation of the two major inputs to dorsal column nuclei cells, the dorsal column fibers and corticofugal fibers, may explain the sensory processing of tactile responses in the dorsal column nuclei. The synaptic complexity of these interactions may be of key functional importance because it could underlie the regulation of sensory infor-

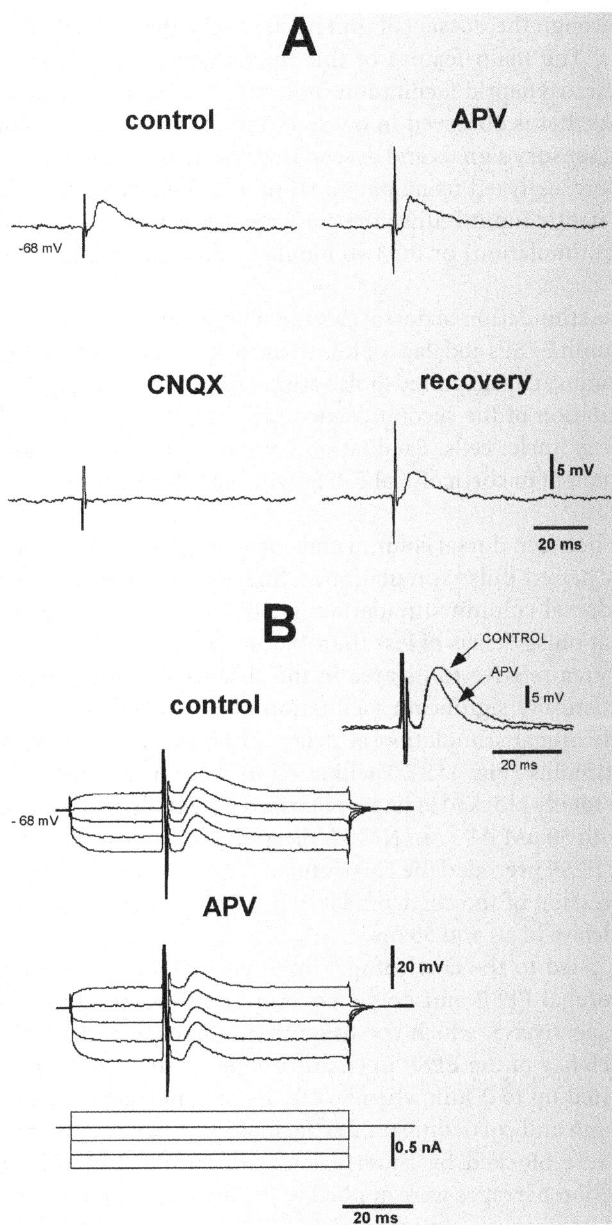

Fig. 9A, B Pharmacological characterization of dorsal column and corticofugal EPSPs. **A** Dorsal column EPSPs (*DC*) were unaffected by APV (50 μM), but were blocked by CNQX (10 μM). The dorsal column EPSP recovered after a 20-min. washout. **B** Corticofugal EPSPs (*CF*) evoked during depolarizing and hyperpolarizing current pulses (current protocol is shown in the lower traces). The late voltage-dependent component was abolished by APV (50 μM). *Inset, right,* shows the block of the late and the reduction of the early component of the corticofugal EPSP by 50 μM APV

mation flow through the dorsal column nuclei to the thalamus by the descending cortical inputs. The main feature of this interaction of excitatory inputs is the homo- and heterosynaptic facilitation induced by activation of descending corticofugal inputs that is observed in anesthetized animals and that could regulate both incoming sensory signals and descending inputs at the cellular level. Synaptic interactions were analyzed using paired stimuli (paired-pulse stimulation) delivered to one synaptic input (either the corticofugal or the dorsal column, termed homosynaptic stimulation) or the two inputs in close succession (heterosynaptic stimulation).

Paired-pulse stimulation of dorsal column fibers elicited a depression of the second dorsal column EPSPs at delays of less than 50 ms in all cells tested (Fig. 10A, C). In contrast, homosynaptic paired-pulse stimulation of corticofugal fibers evoked a modest facilitation of the second corticofugal EPSPs at delays of 10 ms in 60% of dorsal column nuclei cells. Facilitation consisted of a selective increase in the late slow component of corticofugal EPSPs without affecting their peak amplitude (Fig. 10B, C).

Interaction between dorsal column and corticofugal inputs was analyzed using heterosynaptic paired-pulse stimulation at different delays. When the corticofugal preceded the dorsal column stimulation, facilitation of the dorsal column EPSP was observed at pulse delays of less than 100 ms (Fig. 11A). The change in dorsal column EPSP area relative to its area in the absence of corticofugal stimulation revealed a statistically significant facilitation of the dorsal column EPSP by the preceding corticofugal stimulation at delays of between 10 and 50 ms from the corticofugal stimulus (Fig. 11B). Facilitation of the dorsal column EPSP was reduced, or even totally blocked at hyperpolarized potentials <–65 mV or less and by superfusion with 50 µM AP5 (an NMDA receptor antagonist; Fig. 11C). When the dorsal column EPSP preceded the corticofugal EPSP, facilitation did not occur and, instead, a depression of the corticofugal EPSP was observed that was statistically significant at delays of 10 and 50 ms.

Barrages applied to the corticofugal input elicited a short-lasting facilitation of both corticofugal EPSP and dorsal column EPSP (homo- and heterosynaptic facilitation, respectively), which consisted in an increase in the EPSP amplitude and in the efficiency of the EPSP in eliciting action potentials (Fig. 12Aa-Ab and Ba-Bb) that lasted up to 2 min when 50 Hz, 1 s long barrages were applied. Both the dorsal column and corticofugal EPSP facilitations (hetero- and homosynaptic, respectively) were blocked by superfusion with 50 µM AP5 (Fig. 12C). When similar stimulation barrages were applied to the dorsal column input a depression of both dorsal column or corticofugal EPSP areas was observed. Thus, homo- and heterosynaptic facilitation induced by corticofugal stimulation was evoked by activation of NMDA glutamatergic receptors.

Consequently, the presence of an NMDA component in the corticofugal-EPSP, but not in dorsal column-EPSP, suggests relevant functional differences. First, since the NMDA component increases the amplitude and duration of the corticofugal EPSPs evoked at depolarized membrane potentials, it enhances the possibility

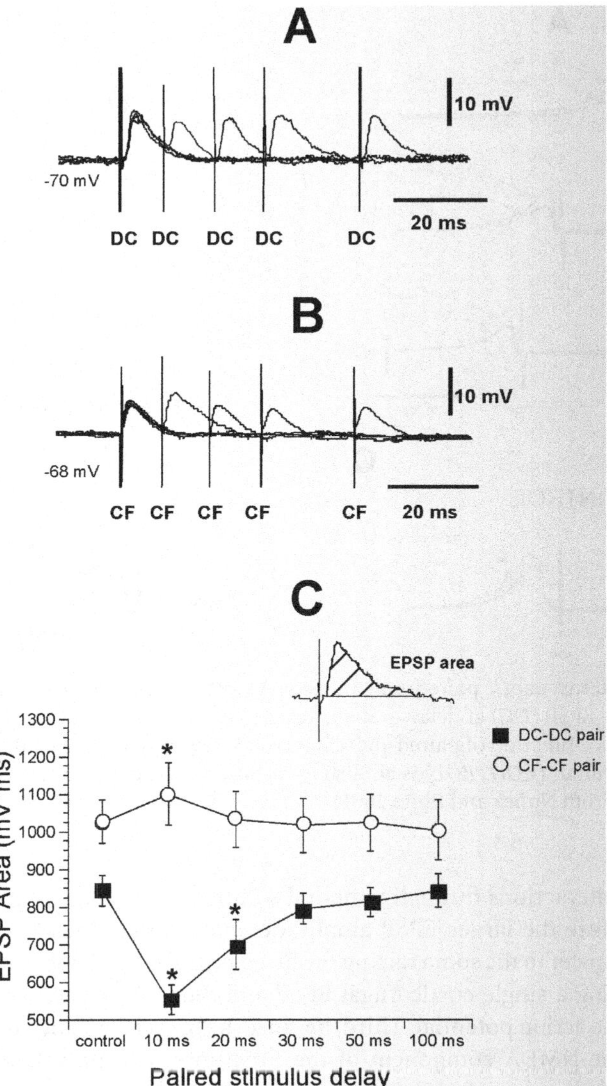

Fig. 10A–C Homosynaptic paired-pulse effects. **A** Superimposed averages ($n = 4$) at different paired-pulse intervals show a depression of the second dorsal column EPSP (*DC*) at delays <50 ms. **B** Superimposed averages ($n = 4$) show small facilitation of the second corticofugal-EPSP (*CF*) at delays of 10–30 ms. **C** Plot of mean EPSP areas of the second EPSP as a function of paired-pulse interval (mean±SEM). *Inset* shows the EPSP area measured. Control values correspond to the area of single EPSPs. (From Nuñez and Buño 2001)

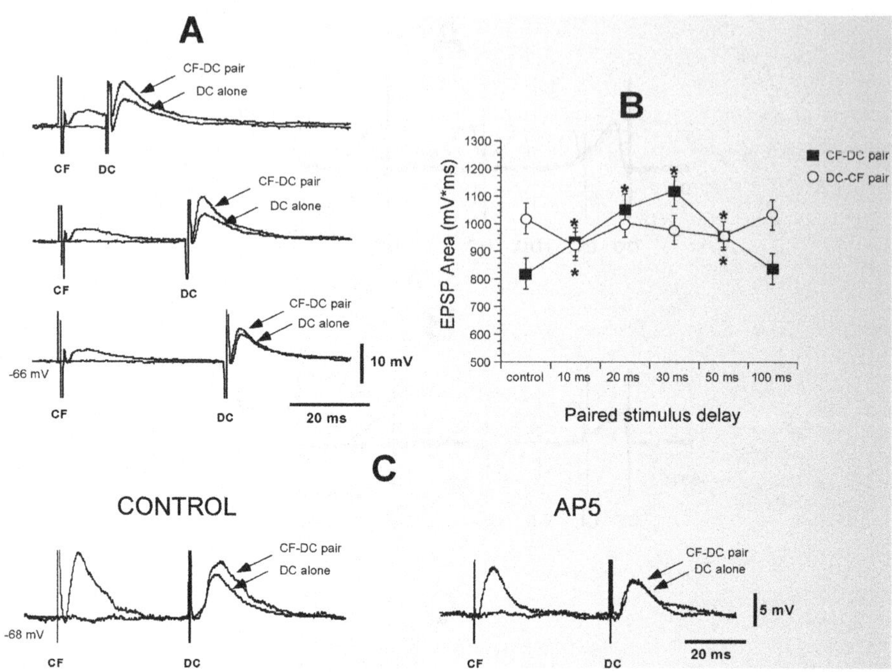

Fig. 11A–C Heterosynaptic paired-pulse effects. **A** Corticofugal (*CF*) stimulation facilitates dorsal column-EPSP (*DC*) at delays <50 ms (averages; $n = 6$). **B** Plot of mean EPSP areas of second EPSP as a function of paired-pulse interval. **C** Heterosynaptic facilitation evoked by corticofugal stimuli (*CONTROL*) is abolish by addition of 50 µM AP5 (an NMDA receptor antagonist). (From Nuñez and Buño 2001)

of synaptic interactions through temporal summation between successive EPSPs. Second, because the larger EPSP amplitude and duration facilitates the passive electronic transfer to the soma raising the firing probability of dorsal column nuclei neurons during a single corticofugal EPSP and thus the possibility of triggering more than one action potential. Third, because the longer sustained depolarization evoked by the NMDA component of the EPSP may activate voltage-gated Ca^{2+} channels and thus increase Ca^{2+} inflow. Finally, and probably most important, Ca^{2+} flowing through NMDA channels may trigger different forms of synaptic plasticity, as has been shown in other systems (see Edmonds et al. 1990; Bliss and Collingridge 1993; Malenka and Nicoll 1993; Clark and Collingridge 1996; Thomson 2000). Facilitation of dorsal column EPSPs was also induced by long-lasting depolarizing current pulses, which may also lead to a substantial increment of intracellular Ca^{2+}, as has been demonstrated in other cell types using Ca^{2+} imaging techniques (Miyakawa et al. 1992; Volgushev et al. 1995). Therefore, Ca^{2+} inflow through NMDA and possibly voltage gated Ca^{2+} channels is probably the key factor in triggering the intracellular mechanisms that generate this type of synaptic facilitation (Bliss and Collingridge 1993).

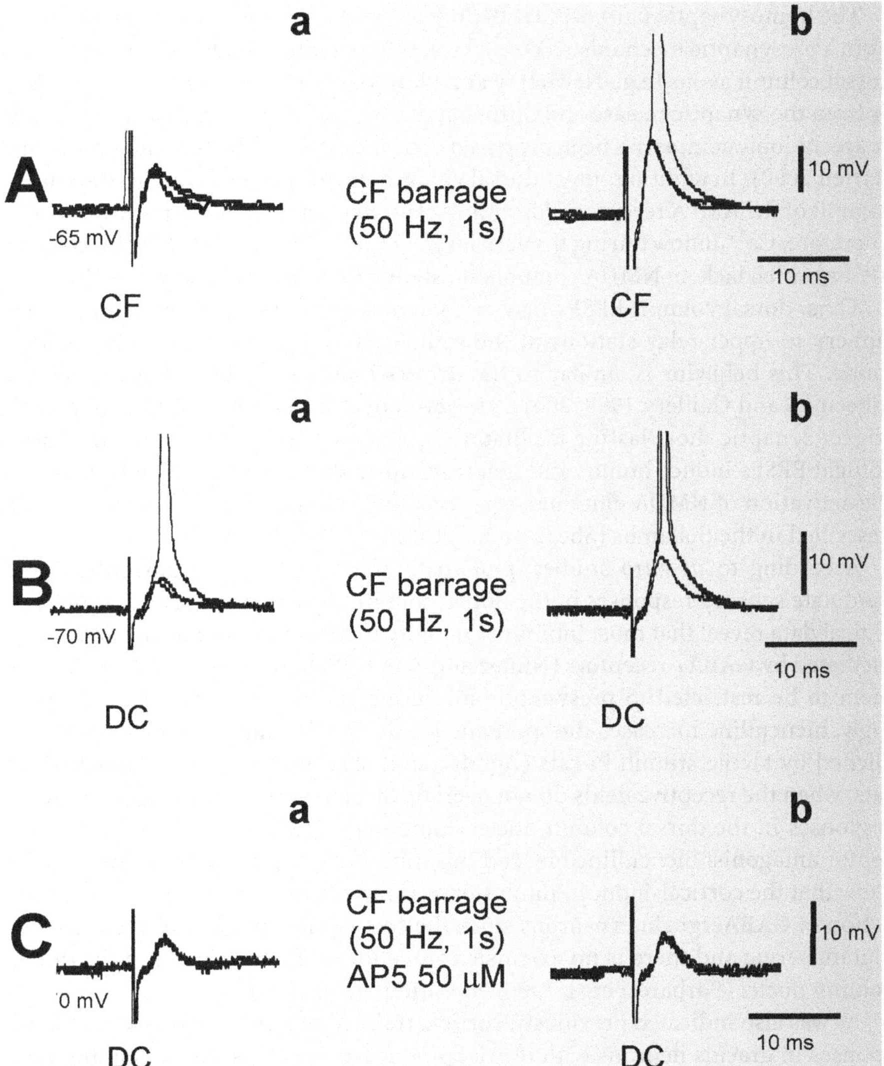

Fig. 12A–C Short-term facilitation evoked by corticofugal barrages. **A** Superimposed traces ($n = 3$) of corticofugal EPSPs (*CF*) before (**a**) and 1 min after (**b**) corticofugal barrage (50 Hz, 1 s). The corticofugal EPSP facilitated and elicited action potentials. Note homosynaptic facilitation. **B** superimposed traces ($n = 3$) of dorsal column EPSPs (*DC*) before (**a**) and 1 min after (**b**) corticofugal barrage (50 Hz, 1 s). The corticofugal barrage facilitated the dorsal column EPSP. Note heterosynaptic facilitation. **C** as in **B** but under 50 µM AP5 blocked the heterosynaptic facilitation. Truncated spikes. (Modified from Nuñez and Buño 2001)

The homosynaptic paired-pulse depression of dorsal column EPSPs may result from a presynaptic mechanism caused by action potential conduction losses in the dorsal column axons (e.g., Newberry and Simmonds 1984; Nuñez and Buño 1999) or from the synaptic release and diffusion of other substances such as nitric oxide or arachidonic acid, as has been proposed to occur in the cerebellum (Reynolds and Hartell 2000). In addition, most dorsal column nuclei neurons contain the GluR2 subunit of the AMPA receptor (Popratiloff et al. 1997) that does not permeate Ca^{2+}. Therefore, Ca^{2+} inflow during the activation of the AMPA-mediated dorsal column EPSPs, which lack an NMDA component, should be negligible (Geiger et al. 1995).

Thus, dorsal column EPSPs may carry out sensory information from the periphery to upper relay stations of the somatosensory pathway in a very precise mode. This behavior is similar to the drivers inputs described in the thalamus (Sherman and Guillery 1998, 2001). However, dorsal column EPSPs are unable to trigger synaptic short-lasting facilitatory processes. On the other hand, the corticofugal EPSPs induce homo- and heterosynaptic short-lasting facilitation due to the activation of NMDA glutamatergic receptors, similar to the modulator inputs described in the thalamus (Sherman and Guillery 1998, 2001).

According to *in vitro* studies, glutamate and GABA neurotransmitters may modulate sensory responses in the dorsal column nuclei. Previous electrophysiological data reveal that most inhibitory activity in dorsal column nuclei neurons is mediated by $GABA_A$ receptors (Nuñez and Buño 1999), whereas $GABA_B$ receptors seem to be restricted to presynaptic inhibition (Deuchars et al. 2000). Accordingly, bicuculline increases the spontaneous firing rate and the number of spikes elicited by tactile stimuli in cats (Aguilar et al. 2003). Moreover, in anesthetized rats, when the receptive fields do not overlap, SI cortex stimulation inhibits tactile responses in the dorsal column nuclei. Iontophoretic ejection of the $GABA_A$ receptor antagonist bicuculline blocked this inhibitory effect (Fig. 13). Thus results show that the cortical-induced inhibition was mostly, if not solely, due to the activation of GABAergic interneurons since the corticofugal projection is exclusively glutamatergic and there is no extrinsic source for GABAergic input to the dorsal column nuclei (Barbaresi et al. 1986; Popratiloff et al. 1996).

As was also indicated previously, cortical train stimulation facilitated tactile responses in gracilis neurons with overlapping receptive fields. After iontophoretic application of APV (50 mM) to the dorsal column nuclei, there was a slight facilitation of tactile responses immediately after cortical stimulation, but this facilitation was not observed 5 min later, in contrast to the results observed in absence of APV (Fig. 14A). Thus, facilitation of tactile responses by corticofugal stimulation in anesthetized animals was caused by activation of NMDA receptors, in agreement with *in vitro* studies. Neurons with nonoverlapping receptive fields with cortical stimulated area were inhibited by cortical train stimulation in control conditions (Fig. 14B). After application of APV to the dorsal column nuclei, cortical stimulation inhibited tactile responses immediately after the stimulation, but tactile responses recovered control values earlier than we expected in control experiments (5 min later). Consequently, results indicate that APV reduces the duration

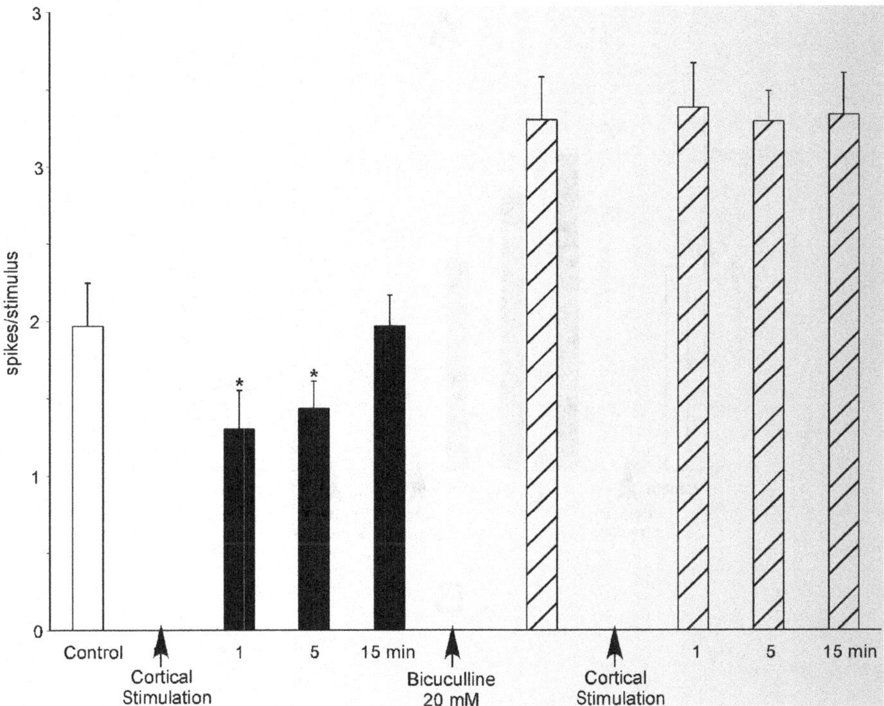

Fig. 13 $GABA_A$ receptor antagonist bicuculline blocks the inhibitory responses evoked by cortical stimulation in anesthetized rats. Plot of tactile responses evoked in seven gracilis neurons with nonoverlapping receptive fields with the cortical stimulated area. Cortical train stimulation evoked a short-lasting inhibition (5 min) of tactile responses. Afterward, iontophoretic application of bicuculline (20 mM, 100–200 nA, 30-s pulse) increased tactile responses. In the presence of bicuculline, cortical stimulation did not inhibit tactile responses. *White bar* control, *black bars* after cortical stimulation, *hatched bars* after bicuculline ejection. (From Malmierca and Nuñez 2004)

of both facilitation and inhibition evoked by corticofugal pathway activation. Thus, NMDA receptors may be located in thalamic projecting neurons as well as in local inhibitory interneurons, affecting their activity.

Also, inhibitory processes could participate in sensory plasticity in dorsal column nuclei by corticofugal stimulation. A paired-pulse paradigm (conditioning and test tactile stimuli) can be used to quantify inhibition in the dorsal column nuclei since the test responses were decreased by previous conditioned stimuli applied with short delays. Thus, two identical tactile stimuli (20 ms at 0.5 Hz) were delivered to the same place in the receptive field and the proportion between test and conditioned responses was calculated at different intervals (30, 50, and 100 ms). In control conditions, the ratio was below one at 30- and 50-ms delays, indicating the existence of an inhibition of test responses, while at 100-ms delays the responses to the conditioning and test stimuli were similar, with a ratio of approx-

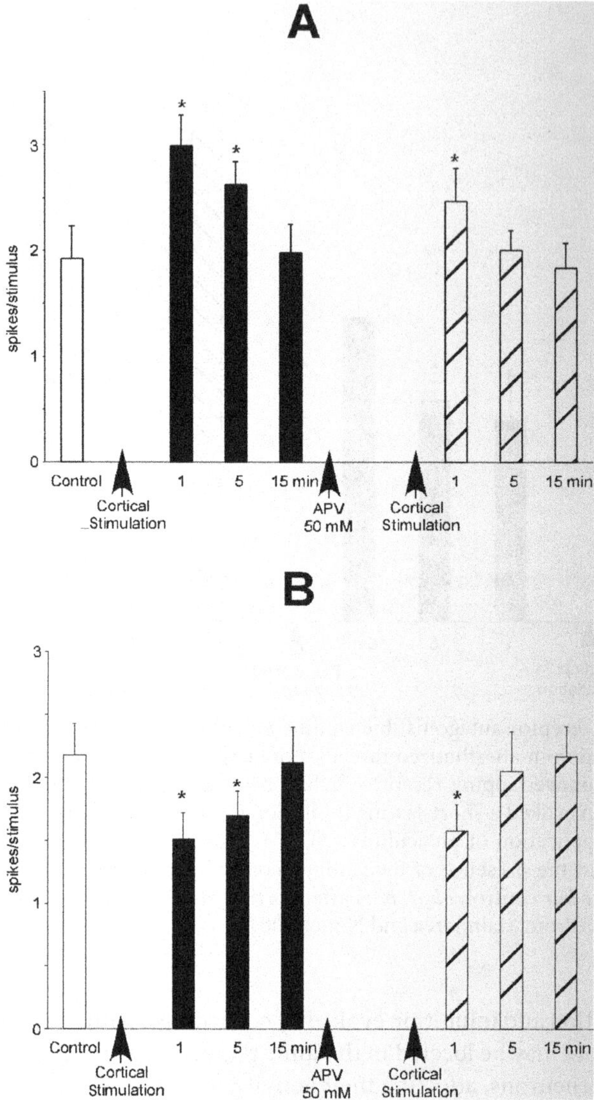

Fig. 14A, B NMDA receptor antagonist APV decreased short-lasting facilitatory and inhibitory responses evoked by cortical stimulation. **A** Plot of tactile responses evoked in six gracilis neurons with overlapping receptive fields. Cortical train stimulation evoked a short-lasting facilitation of tactile responses (up to 5 min). Iontophoretic application of APV (50 mM, 100–200 nA, 30-s pulse) reduced both the amplitude and duration of the facilitation evoked by a new cortical stimulation train. **B** Plot of tactile responses evoked in four gracilis neurons with nonoverlapping receptive fields. Cortical train stimulation evoked a short-lasting inhibition (5 min) in tactile responses. Iontophoretic application of APV (50 mM, 100–200 nA, 30-s pulse) reduced the duration of the inhibition evoked by new cortical stimulation. *White bar,* control; *black bars,* after cortical stimulation; *hatched bars,* after APV ejection. (From Malmierca and Nuñez 2004)

imately 1 (Fig. 15A). In recordings of gracilis neurons with overlapping receptive fields, electrical stimulation of the SI cortex (a train of 0.3-ms pulses at 100 Hz for 500 ms) evoked a decrease in the paired-pulse inhibition at 30- and 50-ms delays, but not at a delay of 100 ms (Fig. 15A). However, in gracilis neurons with nonoverlapping receptive fields, the inhibition elicited by paired-pulse stimulation of the receptive field was not modified by cortical stimulation (Fig. 15B).

These results suggest that during the corticofugal-evoked facilitation in cells with overlapping receptive fields, the inhibition is also decreased. Consequently, the corticofugal pathway may inhibit inhibitory dorsal column nuclei interneurons. Because the corticofugal fibers are glutamatergic, this inhibition of inhibitory interneurons may be caused by the activation of another kind of inhibitory interneuron within the dorsal column nuclei. Recently, Aguilar et al. (2003) demonstrated that GABAergic neurons might be controlled by glycinergic neurons within the dorsal column nuclei, and suggested that when the latter were activated by corticofugal fibers, they would inhibit GABAergic neurons. For this reason, SI cortex may enhance tactile responses by activating non-NMDA and NMDA receptors and by disinhibition of dorsal column nuclei cells, through serial glycinergic–GABAergic interneurons (Aguilar et al. 2003).

5
Functional Considerations

The data summarized above suggest that corticofugal projections modulate sensory processing at all levels of the sensory pathway. The reason for this cortical control is to improve responses to relevant stimuli and to decrease responses from irrelevant stimuli. Thus, the sensory cortex discriminates relevant stimuli from the first relay station of the sensory pathway and favors their transmission (egocentric selection). It has been hypothesized that oscillations could facilitate association between functionally related cells in order to improve neuronal responses. Accordingly, with this role of neuronal oscillations, corticofugal projections also enhance oscillatory activity in order to synchronize neurons located in the same or in different relay stations in order to improve sensory processing. The final consequence of these corticofugal effects is that corticofugal projections contribute to receptive field plasticity and to selective attention processes since they enhance neuronal responses for attentionally relevant stimuli and by suppressing sensory responses of distractive stimuli.

5.1
Corticofugal Modulation of Sensory Transmission: Egocentric Selection

Cortical feedback enhances cortically relevant stimuli, while it decrease other sensory stimuli. This effect has been called egocentric selection (Jen et al. 2002) and play a pivotal role in control the sensory information that reaches the thalamus and cortex.

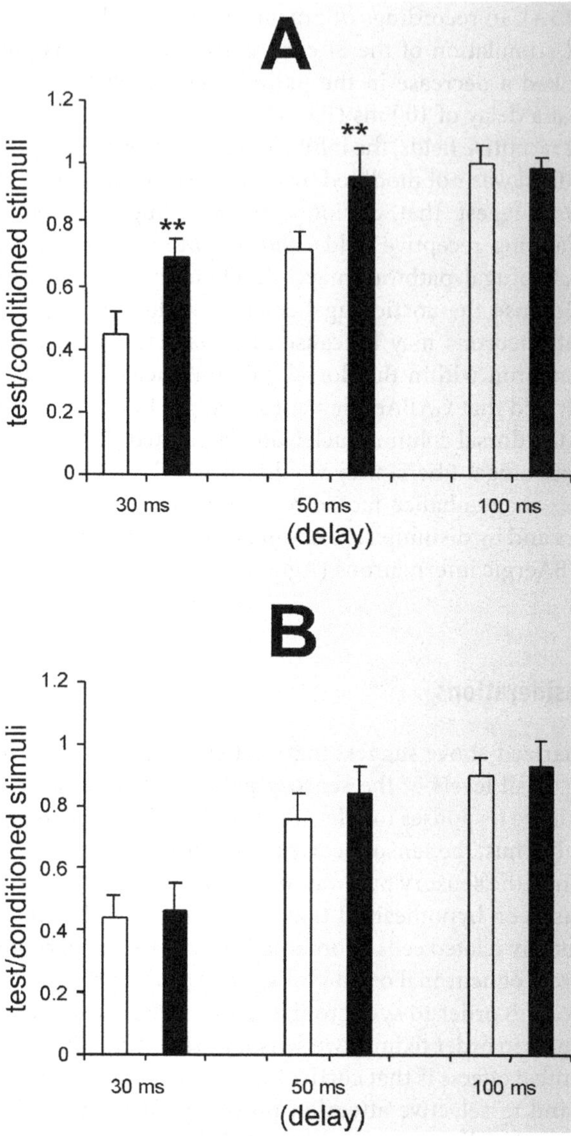

Fig. 15A, B Pair-pulse tactile stimulation in gracilis neurons. **A** Plot of the proportion of responses evoked by two tactile stimuli (20 ms at 0.5 Hz; conditioned and test stimuli) at different delays in gracilis neurons ($n = 14$) with overlapping receptive field as the cortical stimulation site. In the control (*white bars*), test responses were decreased by the conditioned stimuli at 30- and 50-ms delays. One minute after cortical train stimulation (0.3 ms at 100 Hz for 500 ms) the ratio between the test and conditioned responses increased (*black bars*), indicating a reduction in the inhibition. **B** Plot of the ratio of responses evoked by two tactile stimuli in gracilis neurons ($n = 19$) with nonoverlapping receptive fields. Cortical train stimulation did not modify the proportion between test and conditioned responses. (From Malmierca and Nuñez 2004)

Although the basic properties of thalamic receptive fields are undoubtedly established by feed-forward sensory afferents (Sherman and Guillery 1998; Usrey et al. 1999), evidence from the visual, auditory, and somatosensory systems suggests that cortical feedback serves to amplify the effects of sensory stimulation to generate the receptive field of neurons in the sensory pathway. For instance, responses of lateral geniculate neurons to visual stimuli restricted to the functional centre of the receptive field are reduced following cortical inactivation (Przybyszewski et al. 2000), whereas responses to stimuli extending into the surround area display a release of suppression in the absence of cortical feedback (Sillito et al. 1993; Webb et al. 2002). These results suggest that cortical feedback normally serves to enhance the excitatory response of lateral geniculate neurons to stimuli restricted within the classical receptive field, as well as to enhance the suppressive effects of stimuli that extend into the classical surround area. The combination of these two effects could be viewed as complementary mechanisms that serve to increase the spatial filtering properties and/or sharpen the receptive fields of lateral geniculate neurons. Equally, electrical stimulation of the SI cortex reduces the receptive field of dorsal column nuclei (Malmierca and Nuñez 2004; see above).

In the auditory system, cortical feedback can influence the selectivity of medial geniculate neurons in a manner analogous to that of the visual system. For instance, focal activation of auditory cortex in the mustached bat enhances the responsiveness of individual medial geniculate neurons if the best frequency of the thalamic neuron matches the best frequency of the activated region of cortex (Suga et al. 2000; Zhang and Suga 2000). If the best frequencies do not match, then the responsiveness of medial geniculate neurons decreases. Experiments that utilize cortical inactivation find complementary results (Yan and Suga 1996; Zhang et al. 1997). Egocentric selection enhances the filter properties of neurons in the frequency (Zhang et al. 1997; Chowdhury and Suga 2000), amplitude, time (Yan and Suga 1996; Ma and Suga 2001), and spatial (Jen et al. 1998) domains, and augments sensitivity to certain combinations of sounds. Egocentric selection occurs even at the level of the cochlea of the mustached bat (Suga 1984) and probably in humans (Khalfa et al. 2001). Taken together, these results indicate that one role of cortical feedback is to adjust the tuning of subcortical input to the cortex.

In the first relay station of the somatosensory system, corticofugal projections improve tactile responses in the dorsal column nuclei with matching receptive fields and decrease the size of the receptive field (Malmierca and Nuñez 2004). The assumption of a precise cortical control over the dorsal column nuclei is derived not only from quantitative analysis (number of spikes per stimuli), but also from qualitative analysis since receptive field size can be modified by cortical stimulation. The corticofugal pathway restricts receptive field size by reinforcing a functional center of the receptive field and by inhibiting the surrounding area, thus increasing the center–periphery contrast and producing some improvement in the receptive capacity of the somatosensory system (Malmierca and Nuñez 1998, 2004). This functional arrangement probably provides a synaptic mechanism for the enhancement of contrast between relevant and irrelevant sensory signals whereby the corticofugal pathway would selectively facilitate the flow of sensory relevant stimuli to the thalamus and cortex. This improvement could make tactile perception more discriminative.

Less is known about the modulation of suppressive surrounds of somatosensory neurons in the thalamus. Although neurons in area 3b of primate somatosensory cortex are known to have both excitatory and inhibitory regions within their receptive fields (Yan and Suga 1996; Suga et al. 2000), identifying suppressive regions of ventral posterior receptive fields has been more difficult. Nevertheless, studies have shown that the spatial profile of ventral posterior receptive fields can expand following inactivation of SI cortex (Krupa et al. 1999; Ghazanfar et al. 2001). Thus, at least for some ventral posterior neurons, corticothalamic input appears to serve a similar role to that seen in the visual and the auditory system's corticothalamic input serves to sharpen and adjust the profile of thalamic receptive fields.

Therefore, egocentric selection may be a general property of the corticofugal projection in all sensory systems. These observations conceivably result from the existence of a well-organized topographic projection between the sensory cortex and the subcortical relay stations. Consequently, this fine focusing of the corticofugal pathway has a functional significance in spatial discrimination and attention processes.

A similar cortical control of exteroceptive sensory transmission occurred during activation of the motor cortex. Activation of a motor cortical area of a given limb joint selectively facilitates somatosensory responses in dorsal column nuclei neurons that have their receptive field in the body area corresponding to the same joint (Palmeri et al. 1998). This mechanism of facilitation may be similar to what occurred in the dorsal column nuclei with corticofugal modulation of tactile responses by the somatosensory cortex, since facilitation evoked by the motor cortex was reduced when ketamine (an NMDA-receptor antagonist) was used to anesthetize the rats (Giufrida et al. 1985). Therefore, corticofugal projections may also contribute to sensorimotor synchronization.

Topographic maps representing the receptive field can be found in the cortex and in all subcortical relay stations. They are modified by sensory deprivation, injury, and experience even in adult animals (reviews by King 1977; Buonomano and Merzenich 1998). Recently, interest has grown on the reorganization that occurs in the somatosensory system after peripheral nerve injury. Similar to the case in somatosensory cortex, somatosensory thalamus (Florence et al. 2000; Choudhury et al. 2004) and dorsal column nuclei (Pettit and Schwark 1993; Panetsos et al. 1997) also shows massive reorganizational plasticity. Rapid (minutes to hours) reorganization in ventral posterior nucleus can be blocked if the somatosensory cortex is inactivated in the rat (Krupa et al. 1999). The mechanisms proposed to explain the reorganization of cortical somatosensory maps include an increase in the efficacy of relatively weak inputs to the cortex (Merzenich et al. 1984; Jablonska et al. 1995), and local disinhibition (Rasmusson and Turnbull 1983; Hicks et al. 1986; Calford and Tweedale 1991). Similar mechanisms have been proposed in subcortical relay stations such as in the dorsal column nuclei (Pettit and Schwark

1993; Panetsos et al. 1997). The question is raised as to whether new receptive fields result from synaptic changes within the subcortical nucleus or are simply relayed back from a reorganized cortex. Both possibilities may be complementary. Since peripheral deafferentation induces a reorganization of the receptive field in the cortex, this modification may also induce changes in the corticofugal effects in subcortical relay stations. In light of present results, we should consider the likelihood that the corticofugal system may also contribute notably to the plasticity of the receptive field in the subcortical sensory relay stations.

The modulatory effects of corticofugal projections are only possibly because of the existence of a well-organized topographic projection between the cortex and subcortical relay stations. A relevant issue is to determine how this precise pattern of connections is established.

A possible explanation is that this corticofugal topographic projection could be established during development. A fundamental feature of the development of the mammalian brain is the formation of patterned terminations in a target structure by afferents from a source structure. Some of the best-studied examples of developmental pattern formation exist in the visual pathway of higher mammals. In ferrets, for example, retinal axons from the two eyes terminate in eye-specific layers in the lateral geniculate nucleus (Linden et al. 1981), and axons of on-center and off-center retinal ganglion cells from each eye subsequently form on and off sublayers within eye-specific layers (Hahm et al. 1991). Further along in the visual pathway, axons from eye-specific layers of the lateral geniculate nucleus terminate in ocular dominance columns in primary visual cortex (Law et al. 1988), whereas axons of on-center and off-center lateral geniculate nucleus cells terminate in contrast dominance columns within eye-specific columns (Zahs and Stryker 1988; see also Stellwagen and Shatz 2002).

Corticofugal topographic projections may arise as a refinement of initially diffuse and overlapped projections. There are striking differences in the timing of target innervation and rate of elongation of the various populations of corticothalamic axons. Axons of layer V neurons elongate rapidly and enter the thalamus first, whereas those of layer VI neurons grow slowly and enter the thalamus over a prolonged period, subsequent to layer V axons in ferrets (Clascá et al. 1995). Projections from layer V remain massive by the end of the third postnatal week of ferrets. By adulthood, however, they have become drastically reduced although they have apparently reached their appropriate area-specific target nuclei (Clascá et al. 1995). Thus, the most likely interpretation is that some collaterals are removed at later postnatal ages.

Also, a classic example of stereotyped elimination of long axon collaterals occurs during the refinement of subcortical processes arising from layer V cortical neurons (Stanfield et al. 1985). Layer V pyramidal neurons from the motor cortex and the visual cortex initially extend their axons to overlapping targets in the superior colliculus and spinal cord. The neurons from the visual cortex prune their branches from the spinal cord, whereas the neurons from the motor cortex prune their branches from the superior colliculus. Selective elimination appears to be transcriptionally regulated by the homeodomain transcription factor OTX1, because OTX1 translocates to the nucleus of layer V neurons just before pruning (Weimann et al. 1999). In addition, mice with loss-of-function mutations for OTX1 fail to prune away their projections from the spinal cord. The cellular mechanisms that contribute to the stereotyped elimination of layer V axonal projections are unclear, but data suggest that there is an increase in axon degeneration reminiscent of Wallerian degeneration during axon pruning. Also, pruning of long collateral branches could be regulated by extrinsic signals that are received by the axonal branches (for review see Low and Cheng 2005).

The loss of inappropriate synaptic contacts involves activity-dependent competition between the synaptic inputs on single neurons (Shatz and Stryker 1988). The involvement of electrical activity suggests that the pattern of activity may determine whether connections are strengthened or eliminated and that some kind of coincidence detection by the target may be involved. Synchronous firing of two inputs could lead to the strengthening both connections, whereas asynchronous firing could result in strengthening one input but weakening the other. Thus, synchronous activity from *driver* inputs and *modulatory* inputs from the cortex of the same sensory modality, and probably with overlapping receptive fields, may control pruning of corticofugal projections. Consequently, corticofugal pruning during development may contribute to generating the precise topographic pattern of corticofugal projections that is observed in adulthood.

5.2
Corticofugal Modulation of Neuronal Rhythmic Activities and Neuronal Synchronization

One of the most powerful effects of the corticothalamic projection is the facilitation and augmentation of synchronized activity in the interconnected thalamo-cortico-thalamic network. In terms of sensory transmission, the synchronization of stimulus-driven discharges in relay neurons, for example those in the dorsal lateral geniculate nucleus responding to a moving stimulus of a particular orientation (Sillito et al. 1994), or those in the ventral posterior nucleus responding to directionally coded stimuli emanating from a mystacial vibrissa (Ghazanfar et al. 2001; Nicolelis and Shuler 2001), may play a very important role in binding these stimulus features into a single frame of reference for the cerebral cortex. In the visual system, corticothalamic feedback enhances object recognition by increasing the firing synchrony of thalamic relay cells that are coactivated by a moving contour (Sillito et al. 1994; see also Cardin et al. 2005). In the somatosensory thalamus, corticothalamic feedback also enhances sensory responses in topographically aligned barreloids (Temereanca and Simons 2004). Thus, neurons that are spatially distributed, but share common receptive fields may be synchronized by both sensory stimuli and corticofugal projection. The synchronized neurons will induce EPSPs within a narrow temporal window in cells of the next relay station of the sensory pathway, thus increasing their chance of generating spikes. As a result, neuronal synchronization may enhance the contrast between synaptic inputs from synchronized neurons and from nonsynchronized neurons, enhancing the activity of functionally related neurons and filtering out the irrelevant sensory inputs.

In addition, it has been suggested that neuronal oscillatory activity may contribute to spike synchronization of spatially distributed neurons, which would enhance the spatial summation of the synaptic potentials that these neurons evoke in their target cells (Singer 1999; Singer and Gray 1995; Steriade 1993, 2006; Steriade et al. 1993b; Nuñez and Amzica 2004). Neocortical neurons are engaged in cortico-thalamo-cortical loops, which generate different rhythmic activities according to the state of vigilance. These oscillations drive neocortical neurons to fire in a more or less synchronous manner, specific to each behavioral state (Singer 1999; Steriade 1993, 2006; Buzsaki and Draguhn 2004). Rhythmic synchronization may induce synaptic plasticity because modulatory processes at the synapses depend critically on the relative timing of pre- and postsynaptic activation (Huerta and Lisman 1995; Markram et al. 1997). Most of the cortical and subcortical structures of the central nervous system display oscillatory activities in different behavioral conditions, covering a broad range of frequencies (for review see Singer 1993; Steriade et al. 1990, 1993a, b; Steriade 2006). Neuronal oscillatory activities in the central nervous system may be generated by the combination of pacemaker mechanisms in individual cells and reciprocal synaptic interactions among them or within networks of cells that do not oscillate intrinsically but resonate at particular frequencies.

The contribution of oscillations to sensory information processing has been suggested at all levels of the somatosensory system in anesthetized or awake behaving animals (Amassian and Giblin 1974; Murthy and Fetz 1992, 1996; Panetsos et al. 1998; Nuñez et al. 2000). For instance, in the rat trigeminal system, synchronous oscillatory activity at 7–12 Hz appeared in the brainstem, thalamus, and cortex during attentive immobility, and preceded the onset of rhythmic whisker twitching used for tactile exploratory movements (Nicolelis et al. 1995). Thus, synchronous activity in the trigeminal system may not only encode sensory information, but also set up the stage to optimize somatosensory integration.

The reciprocal thalamo-cortico-thalamic loop has been implicated in the generation of several rhythmic activities such as spindles, slow oscillations, and 40-Hz oscillations that characterized the electroencephalogram during different stages of waking-sleep cycle (Steriade et al. 1993a, 1996; Steriade and Timofeev 2003; Nuñez and Amzica 2004; Cardin et al. 2005). The corticothalamic projection also potentiates the genesis of spindle (7–14 Hz) and delta oscillations (Steriade et al. 1991, 1993b; Contreras et al. 1996). During activated states (waking and paradoxical sleep), electroencephalogram recordings are characterized by low-amplitude, high-frequency oscillatory activity in the γ band (30–50 Hz; Steriade et al. 1990, 1996; Llinas et al. 1991). Corticothalamic neurons of cats discharge fast rhythmic spike bursts mainly at 30–40 Hz, suggesting that this activity results in integrated fast oscillations within corticothalamic networks that synchronized thalamic and cortical neuronal populations (Steriade et al. 1998). The major contribution of the corticothalamic projection to the generation of these rhythmic activities is probably the possibility to synchronize cortical neuronal populations according to the convergence–divergence of the corticothalamic and thalamocortical projections. Thus, one of the most powerful effects of the corticothalamic projection is the facilitation and augmentation of synchronized behavior in the interconnected thalamo-cortico-thalamic network.

It has been argued that spatially restricted fast oscillations are an essential step in cortical processing of inputs because they allow formation of temporally coherent, but spatially segregated, clusters of neuronal activity (Singer 1993; Fries et al. 2002). Such temporal coherence has been proposed to establish feature specification (e.g., dot size in the visual field) and cognitive binding (e.g., combination of several features into the perception of a single object) through neuronal synchronization (Singer 1999). In vertebrates, particularly mammals, such temporal binding is thought to occur exclusively at cortical (Singer 1999) or at thalamocortical network levels (Steriade et al. 1996; Llinas et al. 2002). Thalamocortical clusters should stabilize within a few recurrent cycles and have phase-locked coherence that is independent of their location in the cortex. At the level of the whole brain, such thalamocortical organization has been observed in humans using magnetoencephalography (Ribary et al. 1991).

During wakefulness and paradoxical sleep or during tasks involving directed attention, fast rhythms, within β and γ frequency bands (generally 20–60 Hz), in the thalamo-cortico-thalamic network may contribute to the binding process (Gray et al. 1989; Llinas et al. 1991; Fries et al. 2002). These fast rhythms enhance temporal coherence of responses and firing probability of cortical neurons. It appears that spontaneous brain rhythms during different states of vigilance may lead to increased responsiveness and plastic changes in the strength of connections among neurons, a mechanism through which information is stored (Steriade and Timofeev 2003).

Recent data show that synchrony and/or γ-band oscillations are enhanced during the attentional selection of sensory information. Steinmetz et al. (2000) investigated cross-modal attentional shifts in awake monkeys that had to direct attention to either visual or tactile stimuli that were presented simultaneously. For a significant fraction of the neuronal pairs in the second somatosensory area, synchrony depended strongly on the monkey's attention. If the monkey shifted attention to the visual task, temporal correlations typically decreased among somatosensory cells, as compared with task epochs in which attention was not distracted from the somatosensory stimuli. A strong attentional effect on temporal response patterning has also been observed in visual areas in the monkey (Fries et al. 2001). In this study, two stimuli were presented simultaneously on a screen, one inside the receptive fields of the recorded neurons and the other nearby. The animals had to detect subtle changes in one or the other stimulus. If attention was shifted towards the stimulus that was processed by the recorded cells, there was a marked increase in local coherence in the γ band. Evidence for the attentional modulation of neural synchrony is also provided by studies of sensorimotor interactions.

Synchronization between sensory and motor assemblies has been also investigated in awake, behaving cats as they perform a visuomotor coordination task

(Roelfsema et al. 1997). Neural activity was recorded with electrodes that were chronically implanted in various areas of the visual, parietal, and motor cortices. The results show that the synchronization of neural responses occurs not only within the visual system, but also between visual and parietal areas, as well as between parietal and motor cortices.

The topic of this article is beyond the issue of whether or not synchronized fast rhythms in distributed neuronal pools are needed for the representation of multiple facets of the external world into single percepts. However, it is clear that synchronization between cortical and thalamic rhythms may enhance sensory processing. In humans challenged with performance of cognitive tasks, functional MRI showed that the power of fast rhythms (20–35 Hz) increases in the thalamus and occipital cortex during semantic memory recall (Slotnick et al. 2002). This example supports the notion that fast and slow rhythms are elaborated in reciprocal corticothalamic loops and contribute to sensory processing (Steriade et al. 1996).

It has been hypothesized that oscillations in the somatosensory cortex could facilitate an association between functionally related cells (Murthy and Fetz 1993, 1996; Nuñez et al. 2000). This association hypothesis may also be applied to the dorsal column nuclei. Rhythmic activities in the dorsal column nuclei at low (<1 Hz), delta (1–4 Hz), and higher frequency ranges (>4 Hz) have been described in thalamic-projecting cells and presumed interneurons of anesthetized cats and rats (Amassian and Giblin 1974; Mariño et al. 1996, 1999; Panetsos et al. 1998; Nuñez et al. 2000). Low-frequency oscillations appear to originate in the cortex (Mariño et al. 1999; Steriade et al. 1993a, b), whereas higher frequency oscillations seem to arise within the dorsal column nuclei (Panetsos et al. 1998; Nuñez et al. 2000). This hypothesis is in agreement with recent in vitro experiments demonstrating that dorsal column nuclei cells have the capacity to produce rhythmic activity in the absence of both cortical and peripheral inputs (Nuñez and Buño 1999). Panetsos et al. (1998) demonstrated that dorsal column nuclei neurons express coherent oscillatory activity in the 4- to 22-Hz frequency range at single unit, multiunit, and local field potential levels. These oscillations appear spontaneously in presumed local interneurons or during natural sensory stimulation of their receptive fields in thalamic-projecting dorsal column nuclei neurons, which suggests their implication in the processing of the somatosensory information and in its transfer to higher levels of the central nervous system.

This rhythmic activity of dorsal column nuclei is improved when the SI cortex is electrically stimulated. Figure 16 shows the autocorrelograms of a dorsal column nuclei cell in spontaneous condition and during sensory stimulation. Electrical stimulation of a SI cortical matching area induced an increase in the oscillatory activity evoked during tactile stimulation in dorsal column nuclei cells. In contrast, this effect was never observed when a SI cortical nonmatching area was stimulated. Therefore, corticofugal connections improve rhythmic activity in the first relay station of the somatosensory pathway and contribute to synchronize dorsal column nucleus neurons.

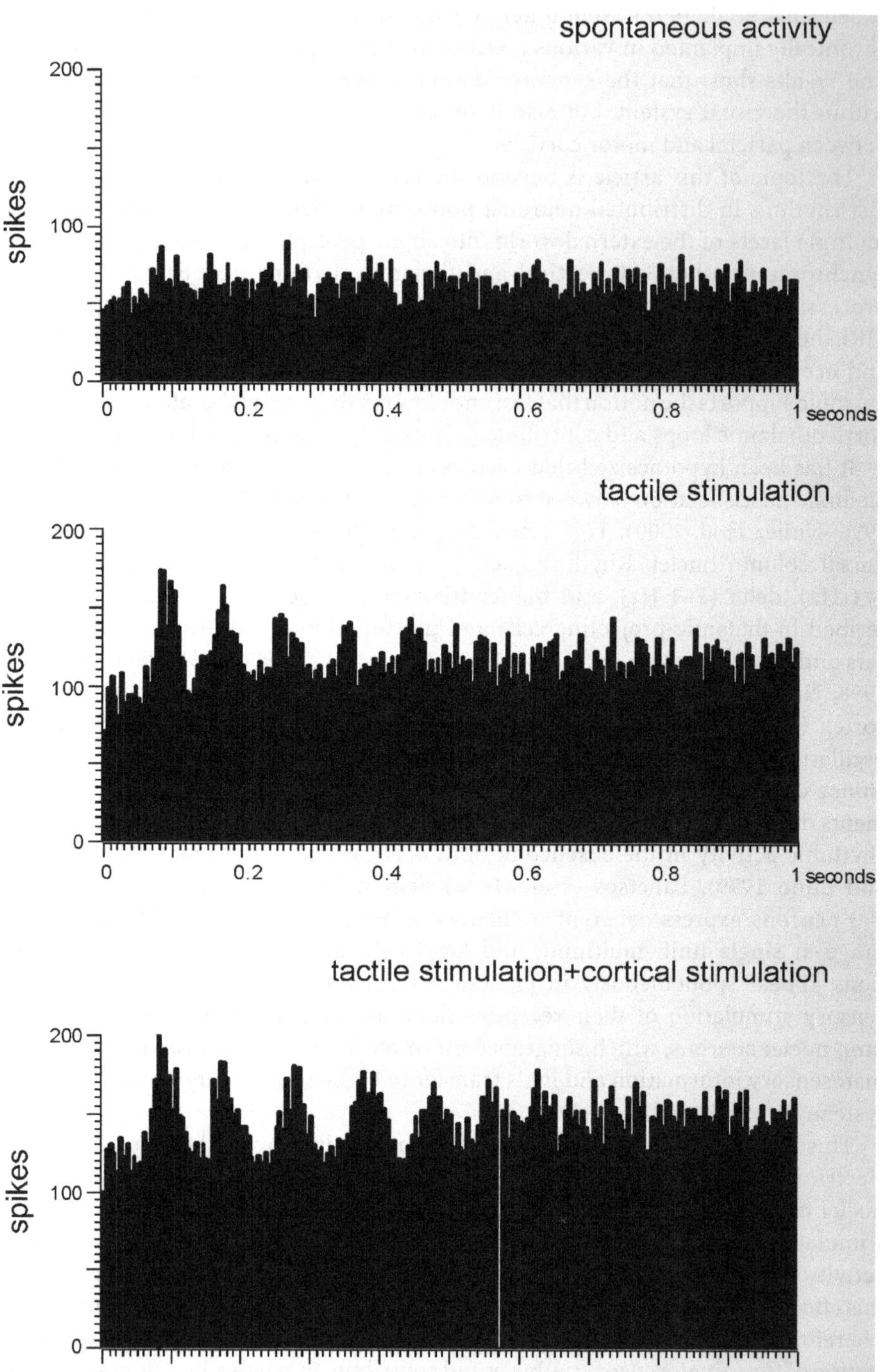

spontaneous activity
spikes
200
100
0
0
0.2
0.4
0.6
0.8
1 seconds
tactile stimulation
spikes
200
100
0
0
0.2
0.4
0.6
0.8
1 seconds
tactile stimulation+cortical stimulation
spikes
200
100
0
0
0.2
0.4
0.6
0.8
1 seconds

Fig. 16 Oscillatory activity of dorsal column nuclei. Autocorrelograms of a representative neuron in the gracilis nucleus. During spontaneous condition, the neuron displayed a non-rhythmic firing pattern (flat autocorrelogram). During tactile stimulation of its receptive field, gracilis neuron showed a rhythmic activity at 10 Hz (peaks in the autocorrelogram indicate a rhythmic activity). One minute after cortical train stimulation (overlapping receptive field condition), tactile stimulation induced a rhythmic activity at 10 Hz as well. However, peaks in the autocorrelogram were more pronounced, indicating that cortical stimulation enhanced rhythmic activity of gracilis cell evoked by sensory stimulation

The data reviewed indicate that corticofugal modulation of sensory processing is, in many instances, associated with modulation of the temporal structure of both ongoing and stimulus-evoked activity, facilitating neuronal oscillations that enhance synchronization of neurons with a common sensory input.

5.3
Contribution of Corticofugal Projections in Attentional Filtering

One suggested role for the corticofugal feedback system is that it may contribute to selective attention. Selective attention is defined as the cognitive function that allows the focusing of processing resources onto the relevant sensory stimuli among the environmental information available at a given moment, while other irrelevant sensory information is largely ignored. Different studies have shown that, in addition to enhancing neuronal processing for attentionally relevant stimuli, selective attention also operates by suppressing the sensory processing of distractive stimuli (Hillyard et al. 1973; Moran et al. 1985; Vidyasagar 1998; Reynolds et al. 1999; Smith et al. 2000; Vanduffel et al. 2000).

The data presented here demonstrated that corticofugal system may mediate attentional modulation of sensory processing in subcortical relay stations. The corticofugal effects on neuronal synchronization, receptive field plasticity, and facilitation of relevant stimuli are examples of neuronal processes that may contribute to selective attention. In the mustached bat, the cochlear best frequency fluctuates during the emission of biosonar pulses (Goldberg and Henson 1998), but it systematically varies with the location of focal cortical activation (Xiao and Suga 2002). In cats, visual attention to a mouse reduces auditory responses of the dorsal cochlear nucleus (Hernández-Peon et al. 1956) and a visual discrimination task reduces auditory nerve responses to clicks (Oatman 1971; Otaman and Anderson 1977). In humans, visual attention reduces auditory nerve responses (Lukas 1980) and sound emissions by the cochlea evoked by a click (Puel et al. 1988). Thus, corticofugal projections act as attentional filters that enhance relevant stimuli and reduce irrelevant stimuli according to the behavioral task in a precise moment.

An important future issue will be to study the relation between different cortical areas in order to favor sensory processing of a specific sensory modality and decrease sensory processing in other sensory pathways.

References

Ahlsen G, Grant K, Lindstrom S (1982) Monosynaptic excitation of principal cells in the lateral geniculate nucleus by corticofugal fibers. Brain Res 234:454–458

Agmon A, Yang LT, Jones EG, O'Dowd DK (1995) Topological precision in the thalamic projection to neonatal mouse barrel cortex. J Neurosci 15:549–561

Aguilar J, Rivadulla C, Soto C, Canedo A (2003) New corticocuneate cellular mechanisms underlying the modulation of cutaneous ascending transmission in anesthetized cats. J Neurophysiol 89:3328–3339

Allegrini PR, Wiessner C (2003) Three-dimensional MRI of cerebral projections in rat brain in vivo after intracortical injection of MnCl2. NMR Biomed 16:252–256

Alloway KD, Hoffer ZS, Hoover JE (2003) Quantitative comparisons of corticothalamic topography within the ventrobasal complex and the posterior nucleus of the rodent thalamus. Brain Res 968:54–68

Amassian VE, Giblin D (1974) Periodic components in steady-state activity of cuneate neurones and their possible role in sensory coding. J Physiol (Lond) 243:353–385

Andersen P, Eccles JC, Oshima T, Schmidt RF (1964a) Mechanisms of synaptic transmission in the cuneate nucleus. J Neurophysiol 27:1096–1116

Andersen P, Eccles JC, Schmidt RF, Yokota T (1964b) Identification of relay cells and interneurons in the cuneate nucleus. J Neurophysiol 27:1080–1095

Andersen P, Junge K, Sveen O (1972) Cortico-fugal facilitation of thalamic transmission. Brain Behav Evol 6:170–184

Andersen RA, Knight PL, Merzenich MM (1980) The thalamocortical and corticothalamic connections of AI, AII and the anterior auditory field (AAF) in the cat: evidence for two largely segregated systems of connections. J Comp Neurol 194:663–701

Andolina IM, Wang W, Jones HE, Sillito AM (2000) Effects of feedback from areas 17, 18 and 19 to the cat LGN. Invest Ophthalmol Vis Sci 41:262

Arcelli P, Frassoni C, Regondi MC, De Biasi S, Spreafico R (1997) GABAergic neurons in mammalian thalamus: a marker of thalamic complexity? Brain Res Bull 42:27–37

Bajo VM, Moore DR (2005) Descending projections from the auditory cortex to the inferior colliculus in the gerbil, Meriones unguiculatus. J Comp Neurol 486:101–116

Baker FH, Malpeli GJ (1977) Effects of cryogenic blockage of visual cortex on the responses of lateral geniculate neurons in the monkey. Exp Brain Res 29:433–444

Bakin JS, Weinberger NM (1996) Induction of physiological memory in the cerebral cortex by stimulation of the nucleus basalis. Proc Natl Acad Sci U S A 93:11219–11224

Banna NR, Jabbur SJ (1989) Neurochemical transmission in the dorsal column nuclei. Somatosen Motor Res 6:237–251

Barbaresi P, Spreafico R, Frassconi C, Rustioni A (1986) GABAergic neurons are present in the dorsal column nuclei but not in the ventroposterior complex of the rat. Brain Res 382:305–326

Bartlett EL, Smith PH (2002) Effects of paired-pulse and repetitive stimulation on neurons in the rat medial geniculate body. Neuroscience 113:957–974

Barlett EL, Stark JM, Guillery RW, Smith PH (2000) Comparison of the fine structure of cortical and collicular terminals in the rat medial geniculate body. Neuroscience 100:811–828

Bentivoglio M, Rustioni A (1986) Corticospinal neurons with branching axons to the dorsal column nuclei in the monkey. J Comp Neurol 253:260–276

Berson DM (1988) Retinal and cortical inputs to cat superior colliculus: composition, convergence and laminar specificity. Prog Brain Res 75:17–26

Bliss TVP, Collingridge GL (1993) A synaptic model of memory: long-term potentiation in the hippocampus. Nature 361:31–39

Bourassa J, Deschênes M (1995) Corticothalamic projections from the primary visual cortex in rats: a single fiber study using biocytin as an anterograde tracer. Neuroscience 66:253–263

Bourassa J, Pinault D, Deschênes M (1995) Corticothalamic projections from the cortical barrel field to the somatosensory thalamus in rats: a single fiber study using biocytin as an anterograde tracer. Eur J Neurosci 7:19–30

Bowling DB, Michael CR (1984) Terminal patterns of single, physiologically characterized optic tract fibers in the cat's lateral geniculate nucleus. J Neurosci 4:198–216

Broman J (1994) Neurotransmitters in subcortical somatosensory pathways. Anat Embryol 189:181–214

Bullier J, Henry GH (1979) Laminar distribution of first-order neurons and afferent terminals in cat striate cortex. J Neurophysiol 42:1271–1281

Buonomano DV, Merzenich, MM (1998) Cortical plasticity: from synapses to maps. Annu Rev Neurosci 21:149–186

Burke W, Dreher B, Wang C (1998) Selective block of conduction in Y optic nerve fibres: significance for the concept of parallel processing. Eur J Neurosci 10:8–19

Buzsáki G, Draguhn A (2004) Neuronal oscillations in cortical networks. Science 304:1926–1929

Calford MB, Tweedale R (1991) Acute changes in cutaneous receptive fields in primary somatosensory cortex after digit denervation in adult flying fox. J Neurophysiol 65:178–187

Canedo A (1997) Primary motor cortex influences on the descending and ascending systems. Prog Neurobiol 51:287–335

Canedo A, Aguilar J (2000) Spatial and cortical influences exerted on cuneothalamic and thalamocortical neurons of the cat. Eur J Neurosci 12:2515–2533

Canedo A, Martinez L, Mariño J (1998) Tonic and bursting activity in the cuneate nucleus of the chloralose-anesthetized cat. Neuroscience 84:603–617

Cardin JA, Palmer LA, Contreras D (2005) Stimulus-dependent (30–50 Hz oscillations in simple and complex fast rhythmic bursting cells in primary visual cortex. J Neurosci 25:5339–5350

Castro-Alamancos MA (2002) Properties of primary (lemniscal) synapses in the ventrobasal thalamus and the relay of high-frequency sensory inputs. Neurophysiol 87:946–953

Catsman-Berrevoets CE, Kuypers HG (1978) Differential laminar distribution of corticothalamic neurons projecting to the VL and the center median. An HRP study in the cynomolgus monkey. Brain Res 154:359–365

Chema S, Whistel BL, Rustioni A (1983) The corticocuneate pathway in the cat: relations among terminal distribution patterns, cytoarchitecture, and single neuron functional properties. Somatosen Res 1:169–205

Chen C, Regehr WG (2003) Presynaptic modulation of the retinogeniculate synapses. J Neurosci 23:3130–3135

Chmielowska J, Carvell GE, Simons DJ (1989) Spatial organization of thalamocortical and corticothalamic projection systems in the rat SmI barrel cortex. J Comp Neurol 285:325–338

Chowdhury SA, Suga N (2000) Reorganization of the frequency map of the auditory cortex evoked by cortical electrical stimulation in the big brown bat. J Neurophysiol 83:1856–1863

Chowdhury SA, Greek KA, Rasmusson DD (2004) Changes in corticothalamic modulation of receptive fields during peripheral injury-induced reorganization. Proc Natl Acad Sci U S A 101:7135–7140

Clascá F, Angelluci A, Sur M (1995) Layer-specific programs of development in neocortical projection neurons. Proc Natl Acad Sci U S A 92:11145–11149

Cole JD, Gordon G (1992) Corticofugal actions on lemniscal neurons of the cuneate, gracilis and lateral cervical nuclei of the cat. Exp Brain Res 90:384–392

Collins CE, Lyon DC, Kaas JH (2005) Distribution across cortical areas of neurons projecting to the superior colliculus in new world monkeys. Anat Rec A Discov Mol Cell Evol Biol 285:619–627

Conti F, De Felipe J, Fariñas I, Manzoni T (1989) Glutamate-positive neurons and axon terminals in cat sensory cortex: a correlative light and electron microscopic study. J Comp Neurol 290:141–153

Contreras D, Destexhe A, Sejnowski TJ, Steriade M (1996) Control of spatiotemporal coherence of a thalamic oscillation by corticothalamic feedback. Science 274:771–774

Cox CL, Sherman SM (2000) Control of dendritic outputs of inhibitory interneurons in the lateral geniculate neurons. Neuron 27:597–610

Cox CL, Huguenard JR, Price DA (1997) Nucleus reticularis neurons mediate diverse inhibitory effects in thalamus. Proc Natl Acad Sci U S A 94:8854–8859

Crabtree JW (1998) Organization in the auditory sector of the cat's thalamic reticular nucleus. J Comp Neurol 390:167–182

Crick F, Koch C (1998) Consciousness and neuroscience. Cereb Cortex 8:97–107

Cudeiro J, Sillito AM (1996) Spatial frequency tuning of orientation-discontinuity-sensitive corticofugal feedback to the cat lateral geniculate nucleus. J Physiol (Lond) 490:481–492

Cudeiro J, Rivadulla C, Grieve KL (2000) Visual response augmentation in cat (and macaque) LGN: potentiation by corticofugally mediated gain control in the temporal domain. Eur J Neurosci 12:1135–1144

De Biasi S, Rustioni A (1990) Ultrastructural immunocytochemical localization of excitatory amino acids in the somatosensory system. J Histochem Cytochem 38:1745–1754

De Biasi S, Vitellaro-Zuccarello L, Bernardi P, Valtschanoff JG, Weinberg R (1994) Ultrastructural and immunocytochemical characterization of terminals of postsynaptic ascending dorsal column fibers in the rat cuneate nucleus. J Comp Neurol 353:109–118

De La Mothe LA, Blumell S, Kajikawa Y, Hackett TA (2006) Thalamic connections of the auditory cortex in marmoset monkeys: core and medial belt regions. J Comp Neurol 496:72–96

Desbois C, Le Bars D, Villanueva L (1999) Organization of cortical projections to the medullary subnucleus reticularis dorsalis: a retrograde and anterograde tracing study in the rat. J Comp Neurol 410:178–196

Deschênes M, Hu B (1990) Electrophysiology and pharmacology of the corticothalamic input to lateral thalamic nuclei: an intracellular study in cat. Eur J Neurosci 2:140–152

Deschênes M, Veinante P, Zhang ZW (1998) The organization of corticothalamic projections: reciprocity versus parity. Brain Res Rev 28:286–308

Deuchars SA, Trippenbach T, Spyer M (2000) Dorsal column nuclei neurons recorded in a brain stem-spinal cord projection: characteristics and their responses to dorsal root stimulation. J Neurophysiol 84:1361–1368

Diamond IT, Jones EG, Powell TPS (1969) The projection of the auditory cortex upon the diencephalon and brainstem in the cat. Brain Res 15:305–340

Diamond ME, Armstrong-James M, Budway MJ, Ebner FF (1992) Somatic sensory responses in the rostral sector of the posterior group Pom and in the ventral posterior medial nucleus VPM of the rat thalamus: dependence on the barrel field cortex. J Comp Neurol 319:66–84

Donoghue JP, Wenthold RJ, Altschuler RA (1985) Localization of glutaminase-like and aspartate aminotransferase-like immunoreactivity in neurons of cerebral neocortex. J Neurosci 5:2597–2608

Doucet JR, Rose L, Ryugo DK (2002) The cellular origin of corticofugal projections to the superior olivary complex in the rat. Brain Res 925:28–41

Doucet JR, Molavi DL, Ryugo DK (2003) The source of corticocollicular and corticobulbar projections in area Te1 of the rat. Exp Brain Res 153:461–466

Eaton SA, Salt TE (1996) Role of N-methyl-D-aspartate and metabotropic glutamate receptors in corticothalamic excitatory postsynaptic potentials in vivo. Neuroscience 73:1–5

Edmonds B, Klein M, Dale N, Kandel ER (1990) Contributions of two types of calcium channels to synaptic transmission and plasticity. Science 250:1142–1147

Ergenzinger ER, Glasier MM, Hahm JO, Pons TP (1998) Cortically induced thalamic plasticity in the primate somatosensory system. Nat Neurosci 1:226–229

Erisir A, Van Horn SC, Sherman SM (1997) Relative numbers of cortical and brainstem inputs to the lateral geniculate nucleus. Proc Natl Acad Sci U S A 94:1517–1520

Fagg GE, Foster AC (1983) Amino acid neurotransmitters and their pathways in the mammalian central nervous system. Neuroscience 9:701–719

Faye-Lund H (1985) The neocortical projection to the inferior colliculus in the albino rat. Anat. Embryol Berl 173:53–70

Feig S, Harting JK (1998) Corticocortical communication via the thalamus: ultrastructural studies of corticothalamic projections from area 17 to the lateral posterior nucleus of the cat and inferior pulvinar nucleus of the owl monkey. J Comp Neurol 395:281–295

Feliciano M, Potashner SJ (1995) Evidence for a glutamatergic pathway from the guinea pig auditory cortex to the inferior colliculus. J Neurochem 65:1348–1357

Ferster D (1988) Spatially opponent excitation and inhibition in simple cells of the cat visual cortex. J Neurosci 8:1172–1180

Florence SL, Hackett TA, Strata F (2000) Thalamic and cortical contributions to neural plasticity after limb amputation. J Neurophysiol 83:3154–3159

Fries P, Reynolds JH, Rorie AE, Desimone R (2001) Modulation of oscillatory neuronal synchronization by selective visual attention. Science 291:1560–1563

Fries P, Schröder JH, Roelfsema PR, Singer W, Engel AK (2002) Oscillatory neuronal synchronization in primary visual cortex as a correlate of stimulus selection. J Neurosci 22:3739–3754

Galuske RA, Schmidt KE, Goebel R, Lomber SG, Payne BR (2002) The role of feedback in shaping neural representations in cat visual cortex. Proc Natl Acad Sci U S A 99:17083–17088

Games KD, Winer JA (1988) Layer V in rat auditory cortex: projections to the inferior colliculus and contralateral cortex. Hear Res 34:1–25

Gao E, Suga N (2000) Experience-dependent plasticity in the auditory cortex and the inferior colliculus of bats: role of the corticofugal system. Proc Natl Acad Sci U S A 97:8081–8086

Geiger JRP, Melcher T, Koh DS, Sakmann B, Seeburg PH, Jonas P, Mpnyer H (1995) Relative abundance of subunit mRNAs determine gating and Ca^{2+} permeability of AMPA receptors in principal neurons and interneurons in rat CNS. Neuron 15:193–204

Ghazanfar AA, Krupa DJ, Nicolelis MA (2001) Role of cortical feedback in the receptive field structure and nonlinear response properties of somatosensory thalamic neurons. Exp Brain Res 141:88–100

Geiger JRP, Melcher T, Koh DS, Sakmann B, Seeburg PH, Jonas P, Monyer H (1995) Relative abundance of subunit mRNAs determines gating and Ca^{2+} permeability of AMPA receptors in principal neurons and interneurons in rat CNS. Neuron 15:193–204

Geisert EE, Langsetmo A, Spear FD (1981) Influence of the corticogeniculate pathway on response properties of cat lateral geniculate neurons. Brain Res 208:409–415

Gilbert CD (1977) Laminar differences in receptive field properties of cells in cat primary visual cortex. J Physiol. (Lond) 268:391–421

Gilbert CD, Kelly JP (1975) The projections of cells in different layers of the cat's visual cortex. J Comp Neurol 163:81–105

Gilbert CD, Wiesel TN (1979) Morphology and intracortical projections of functionally characterised neurones in the cat visual cortex. Nature 280:120–125

Giuffrida R, Li Volsi G, Panto MR, Perciavalle V, Urbano A (1983) Pyramidal influences on ventral thalamic nuclei in the cat. Brain Res 279:254–257

Giuffrida R, Sanderson P, Sapienza S (1985) Effect of microstimulation of movement evoking cortical foci on the activity of neurons of the dorsal column nuclei. Somatosensory Res 3:37–47

Godwin DW, Van Horn SC, Eriir A, Sesma M, Romano C, Sherman SM (1996) Ultrastructural localization suggests that retinal and cortical inputs access different metabotropic glutamate receptors in the lateral geniculate nucleus. J Neurosci 16:8181–8192

Goldberg RL, Henson OWJr (1998) Changes in cochlear mechanics during vocalization: evidence for a phasic medial efferent effect. Hear Res 122:71–81

Gordon G, Jukes GM (1964) Descending influences on the exteroceptive organization of the cat's nucleus gracilis. J Physiol (Lond) 173:291–319

Gray CM, König P, Engel AK, Singer W (1989) Oscillatory responses in cat visual cortex exhibit inter-columnar synchronization which reflects global stimulus properties. Nature 338:334–337

Grieve KL, Sillito AM (1991) The length summation properties of layer VI cells in the visual cortex and hypercomplex cell end zone inhibition. Exp Brain Res 84:319–325

Grieve KL, Sillito AM (1995a) Differential properties of cells in the feline primary visual cortex providing the corticofugal feedback to the lateral geniculate nucleus and visual claustrum. J Neurosci 15:4868–4874

Grieve KL, Sillito AM (1995b) Non-length-tuned cells in layers II/III and IV of the visual cortex: the effect of blockade of layer VI on responses to stimuli of different lengths. Exp Brain Res 104:12–20

Guillery RW (1966) A study of Golgi preparations from the dorsal lateral geniculate nucleus of the cat. J Comp Neurol 128:21–50

Guillery RW, Feig SL, Van Lieshout DP (2001) Connections of higher order visual relays in the thalamus: a study of corticothalamic pathways in cats. J Comp Neurol 438:66–85

Hahm J-O, Langdon RB, Sur M (1991) Disruption of retinogeniculate afferent segregation by antagonists to NMDA receptors. Nature 351:568–570

Harting JK, Updyke BV, Van Lieshout DP (1992) Corticotectal projections in the cat: anterograde transport studies of twenty-five cortical areas. J Comp Neurol 324:379–414

Harvey AR (1980) A physiological analysis of subcortical and commissural projections of areas 17 and 18 of the cat. J Physiol (Lond) 302:507–534

Hashemi-Nezhad M, Wang C, Burke W, Dreher B (2003) Area 21a of cat visual cortex strongly modulates neuronal activities in the superior colliculus. J Physiol (Lond) 550:535–552

Hazama M, Kimura A, Donishi T, Sakoda T, Tamai Y (2004) Topography of corticothalamic projections from the auditory cortex of the rat. Neuroscience 124:655–667

He J (1997) Modulatory effects of regional cortical activation on the onset responses of the cat medial geniculate neurons. J Neurophysiol 77:896–908

He J (2003) Corticofugal modulation of the auditory thalamus. Exp Brain Res 153:579–590

He J, Yu YQ, Xiong Y, Hashikawa T, Chan YS (2002) Modulatory effect of cortical activation on the lemniscal auditory thalamus of the guinea pig. J Neurophysiol 88:1040–1050

Hefti BJ, Smith PH (2000) Anatomy, physiology, and synaptic responses of rat layer V auditory cortical cells and effects of intracellular GABAA blockade. J Neurophysiol 83:2626–2638

Hernández-Peon R, Scherrer H, Jouvet M (1956) Modification of electric activity in cochlear nucleus during attention in unanesthetized cats. Science 123:331–332

Hicks TP, Metherate R, Landry P, Dykes RW (1986) Bicuculline-induced alterations of response properties in functionally identified ventroposterior thalamic neurones. Exp Brain Res 63:248–264

Hillyard SA, Hink RF, Schwent VL, Picton TW (1973) Electrical signs of selective attention in the human brain. Science 12:177–180

Hirsch JA, Alonso JM, Reid RC, Martinez LM (1998) Synaptic integration in striate cortical simple cells. J Neurosci 18:9517–9528

Hoffmann KP (1973) Conduction velocity in pathways from retina to superior colliculus in the cat: a correlation with receptive-field properties. J Neurophysiol 36:409–424

Hoogland PV, Welker A, Van der Loos H (1987) Organization of the projections from barrel cortex to thalamus in mice studied with Phaseolus fulgaris-leucoagglutinin and HRP. Exp Brain Res 68:73–87

Hoogland PV, Wouterlood FG, Welker E, Van der Loos H (1991) Ultrastructure of giants and small thalamic terminals of cortical origin: a study of the projections from the barrel cortex in mice using Phaseolus fulgaris-leucoagglutinin PHA-L. Exp Brain Res 87:159–172

Hu B (1993) Membrane potential oscillations and corticothalamic connectivity in rat associational thalamic neurons in vitro. Acta Physiol Scand 148:109–113

Hubel DH, Wiesel TN (1962) Receptive fields, binocular interaction and functional architecture in the cat's visual cortex. J Physiol (Lond) 160:106–154

Huerta PT, Lisman JE (1995) Bidirectional synaptic plasticity induced by a single burst during cholinergic theta oscillations in CA1 in vitro. Neuron 15:1053–1063

Huffman RF, Henson OWJr (1990) The descending auditory pathway and acousticomotor systems: connections with the inferior colliculus. Brain Res Rev 15:295–323

Irvin GE, Casagrande VA, Norton TT (1993) Center/surround relationships of magnocellular, parvocellular, and koniocellular relay cells in primate lateral geniculate nucleus. Vis Neurosci 10:363–73

Jabbur SJ, Towe AL (1961) Cortical excitation of neurons in dorsal column nuclei of cat, including an analysis of pathways. J Neurophysiol 24:499–509

Jablonska B, Gierdalski M, Siucinska E, Skangiel KJ, Kossut M (1995) Partial blocking of NMDA receptors restricts plastic changes in adult mouse barrel cortex. Behav Brain Res 66:207–216

Jacquin MF, Wiegand MR, Renehan WE (1990) Structure-function relationships in rat brain stem subnucleus interpolaris. VIII. Cortical inputs. J Neurophysiol 64:3–27

Jen PH, Chen QC, Sun X (1998) Corticofugal regulation of auditory sensitivity in the bat inferior colliculus. J Comp Physiol A 183:683–697

Jen PHS, Zhou X, Zhang J, Sun X (2002) Brief and short-term corticofugal modulation of acoustic signal processing in the bat midbrain. Hear Res 168:196–207

Jones EG (1975) Lamination and differential distribution of thalamic afferents within the sensory-motor cortex of the squirrel monkey. J Comp Neurol 160:167–203

Jones EG (1985) The Thalamus. Plenum Press, New York

Jones EG, Powell TP (1969) The cortical projection of the ventroposterior nucleus of the thalamus in the cat. Brain Res 13:298–318

Jones EG, Wise SP (1977) Size, laminar and columnar distribution of efferent cells in the sensory-motor cortex of monkeys. J Comp Neurol 175:391–438

Jones EG, Wise SP, Coulter JD (1979) Differential thalamic relationship of sensory motor and parietal cortical fields in monkeys. J Comp Neurol 183:833–882

Jones HE, Andolina IM, Oakely NM, Murphy PC, Sillito AM (2000) Spatial summation in lateral geniculate nucleus and visual cortex. Exp Brain Res 135:279–284

Jung SC, Shin HC (2002) Reversible changes of presumable synaptic connections between primary somatosensory cortex and ventral posterior lateral thalamus of rats during temporary deafferentation. Neurosci Lett 331:111–114

Kalil RE, Chase R (1970) Corticofugal influence on activity of lateral geniculate neurons in the cat. J Neurophysiol 33:459–474

Katz LC (1987) Local circuitry of identified projection neurons in cat visual cortex brain slices. J Neurosci 7:1223–1249

Kawamura K, Konno T (1979) Various types of corticotectal neurons of cats as demonstrated by means of retrograde axonal transport of horseradish peroxidase. Exp Brain Res 35:161–175

Kelly JP, Wong D (1981) Laminar connections of the cat's auditory cortex. Brain Res 212:1–15

Khalfa S, Bougeard R, Morand N, Veuillet E, Isnard J, Guenot M, Ryvlin P, Fischer C, Collet L (2001) Evidence of peripheral auditory activity modulation by the auditory cortex in humans. Neuroscience 104:347–358

Kilgard MP, Merzenich MM (1998) Plasticity of temporal information processing in the primary auditory cortex. Nat Neurosci 1:727–731

Killackey HP, Koralek KA, Chiaia NL, Rhodes RW (1989) Laminar and areal differences in the origin of the subcortical projection neurons of the rat somatosensory cortex. J Comp Neurol 282:428–445

Kimura A, Donishi T, Sakoda T, Hazama M, Tamai Y (2003) Auditory thalamic nuclei projections to the temporal cortex in the rat. Neuroscience 117:1003–1016

Kimura A, Donishi T, Okamoto K, Tamai Y (2004) Efferent connections of "posterodorsal" auditory area in the rat cortex: implications for auditory spatial processing. Neuroscience 128:399–419

Kimura A, Donishi T, Okamoto K, Tamai Y (2005) Topography of projections from the primary and non-primary auditory cortical areas to the medial geniculate body and thalamic reticular nucleus in the rat. Neuroscience 135:1325–1342

Komura Y, Tamura R, Uwano T, Nishijo H, Kaga K, Ono T (2001) Retrospective and prospective coding for predicted reward in the sensory thalamus. Nature 412:546–549

King AJ (1997) Sensory processing: signal selection by cortical feedback. Curr Biol 7:R85–R88

Krnjevic K, Phillis JW (1963) Actions of certain amines on cerebral cortical neurones. Br J Pharmacol Chemother 20:471–490

Krupa DJ, Ghazanfar AA, Nicolelis MA (1999) Immediate thalamic sensory plasticity depends on corticothalamic feedback. Proc Natl Acad Sci U S A 96:8200–8205

Kus L, Saxon D, Beitz AJ (1995) NMDA R1 mRNA distribution in motor and thalamic projecting sensory neurons in the rat spinal cord and brain stem. Neurosci Lett 25:201–204

Kuypers HG, Lawrence DG (1967) Cortical projections to the red nucleus and the brain stem in the Rhesus monkey. Brain Res 4:151–188

Kuypers HGJM, Tuerk JD (1964) The distribution of the cortical fibres within the nuclei cuneatus and gracilis in the cat. J Anat (Lond) 98:143–162

Land PW, Buffer SAJr, Yaskosky JD (1995) Barreloids in adult rat thalamus: three-dimensional architecture and relationship to somatosensory cortical barrels. J Comp Neurol 355:573–588

Landry P, Dykes RW (1985) Identification of two populations of corticothalamic neurons in cat primary somatosensory cortex. Exp Brain Res 60:289–298

Law MI, Zahs KR, Stryker MP (1988) Organization of primary visual cortex (area 17) in the ferret. J Comp Neurol 278:157–180

Lent R (1982) The organization of subcortical projections of the hamster's visual cortex. J Comp Neurol 206:227–242

Levesque M, Parent A (1998) Axonal arborization of corticostriatal and corticothalamic fibers arising from prelimbic cortex in the rat. Cereb Cortex 8:602–613

Levesque M, Gagnon S, Parent A, Deschênes M (1996) Axonal arborizations of corticostriatal and corticothalamic fibers arising from the second somatosensory area in the rat. Cereb Cortex 6:759–770

Lewitt M, Carreras M, Liu CN, Chambers WW (1964) Pyramidal and extrapyramidal modulation of somatosensory activity in gracilis and cuneate nuclei. Archs Ital Biol 102:197–229

Li J, Guido W, Bickford ME (2003) Two distinct types of corticothalamic EPSPs and their contribution to short-term synaptic plasticity. J Neurophysiol 90:3429–3440

Linden DC, Guillery RW, Cucchiaro J (1981) The dorsal lateral geniculate nucleus of the normal ferret and its postnatal development. J Comp Neurol 203:189–211

Liu XB, Jones EG (1996) Localization of alpha type II calcium calmodulin-dependent protein kinase at glutamatergic but not gamma-aminobutyric acid (GABAergic) synapses in thalamus and cerebral cortex. Proc Natl Acad Sci U S A 93:7332–7336

Liu XB, Honda CN, Jones EG (1995) Distribution of four types of synapse on physiologically identified relay neurons in the ventral posterior thalamic nucleus of the cat. J Comp Neurol 352:69–91

Livingstone MS, Hubel DH (1984) Anatomy and physiology of a colour system in the primate visual cortex. J Neurosci 4:309–356

Llinas R, Grace AA, Yarom Y (1991) In vitro neurons in mammalian cortical layer 4 exhibit intrinsic oscillatory activity in the 10 to 50 Hz frequency. Proc Natl Acad Sci U S A 88:897–901

Llinas R, Leznik E, Urbano FJ (2002) Temporal binding via cortical coincidence detection of specific and nonspecific thalamocortical inputs: a voltage-dependent dye-imaging study in mouse brain slices. Proc Nat Acad Sci U S A 99:449–454

Low LK, Cheng H-J (2005) A little nip and tuck: axon refinement during development and axonal injury. Curr Opin Neurobiol 15:549–556

Lue JH, Jiang SY, Shieh JY, Wen CY (1996) The synaptic interrelationships between primary afferent terminals, cuneothalamic relay neurons and GABA-immunoreactive boutons in the rat cuneate nucleus. J Neurosci Res 24:363–371

Lukas JH (1980) Human auditory attention: the olivocochlear bundle may function as a peripheral filter. Psychophysiology 17:444–452

Lund JS, Lund RD, Hendrickson AE, Bunt AH, Fuchs AF (1975) The origin of efferent pathways from the primary visual cortex, area 17, of the macaque monkey as shown by retrograde transport of horseradish peroxidase. J Comp Neurol 164:287–303

Ma X, Suga N (2001) Plasticity of bat's central auditory system evoked by focal electric stimulation of auditory and/or somatosensory cortices. J Neurophysiol 85:1078–1087

Maioli MG, Domeniconi R, Squatrito S, Riva-Sanseverino E (1992) Projections from cortical visual areas of the superior temporal sulcus to the superior colliculus, in macaque monkeys. Arch Ital Biol 130:157–166

Malenka RC, Nicoll RA (1993) NMDA-receptor-dependent synaptic plasticity: multiple forms and mechanisms. Trends Neurosci 16:521–526

Malmierca E, Nuñez A (1998) Corticofugal action on somatosensory response properties of rat nucleus gracilis cells. Brain Res 810:172–180

Malmierca E, Nuñez A (2004) Primary somatosensory cortex modulation of tactile responses in nucleus gracilis cells of rats. Eur J Neurosci 19:1572–1580

Mariño J, Martinez L, Canedo A (1996) Coupled slow and delta oscillations between cuneothalamic and thalamocortical neurons in the chloralose anesthetized cat. Neurosci Lett 219:107–110

Mariño J, Aguilar J, Canedo A (1999) Cortico-subcortical synchronization in the chloralose-anesthetized cat. Neuroscience 93:409–411

Mariño J, Canedo A, Aguilar J (2000) Sensorimotor cortical influences on cuneate nucleus rhythmic activity in the anesthetized cat. Neuroscience 95:657–673

Markram H, Lübke J, Frotscher M, Sakmann B (1997) Regulation of synaptic efficacy by coincidence of postsynaptic APs and EPSPs. Science 275:213–215

Marrocco RT, McClurkin JW, Alkire MT (1996) The influence of the visual cortex on the spatiotemporal response properties of lateral geniculate nucleus cells. Brain Res 737:110–118

Martin LJ, Blackstone CD, Huganir RL, Price DL (1992) Cellular localization of a metabotropic glutamate receptor in rat brain. Neuron 9:259–270

Martinez L, Lamas JA, Canedo A (1995) Pyramidal tract and corticospinal neurons with branching axons to the dorsal column nuclei of the cat. Neuroscience 68:195–206

Martinez-Lorenzana G, Machin R, Avendaño C (2001) Define segregation of cortical neurons projecting to the dorsal column nuclei in the rat. Neuroreport 12:413–416

Mayner L, Kaas JH (1986) Thalamic projections from electrophysiologically defined sites of body surface representations in areas 3b and 1 of somatosensory cortex of Cebus monkeys. Somatosens Res 4:13–29

McComas AJ, Wilson P (1968) An investigation of pyramidal tract cells in the somatosensory cortex of the rat. J Physiol (Lond) 194:271–288

McCormick DA, von Krosigk M (1992) Corticothalamic activation modulates thalamic firing through glutamate metabotropic receptors. Proc Natl Acad Sci U S A 89:1774–1778

McCourt ME, Boyapati J, Henry GH (1986) Layering in lamina 6 of cat striate cortex. Brain Res 364:181–185

McHaffie JG, Thomson CM, Stein BE (2001) Corticotectal and corticostriatal projections from the frontal eye fields of the cat: an anatomical examination using WGA-HRP. Somatosens Mot Res 18:117–130

Meredith MA, Clemo HR, Stein BE (1991) Somatotopic component of the multisensory map in the deep laminae of the cat superior colliculus. J Comp Neurol 312:353–370

Merzenich MM, Nelson RJ, Stryker MP, Cynader MS, Schoppmann A, Zook JM (1984) Somatosensory cortical map changes following digit amputation in adult monkeys. J Comp Neurol 224:591–604

Mitani A, Shimokouchi M, Nomura S (1983) Effects of stimulation of the primary auditory cortex upon colliculogeniculate neurons in the inferior colliculus of the cat. Neurosci Lett 42:185–189

Miyakawa H, Lev Ram V, Lasser-Ross N, Ross WN (1992) Calcium transients evoked by climbing fibre and parallel fibre synaptic inputs in guinea pig cerebellar Purkinge neurons. J Neurophysiol 68:1178–1189

Molinari M, Minciacchi D, Bentivoglio M, Macchi G (1985) Efferent fibers from the motor cortex terminate bilaterally in the thalamus of rats and cats. Exp Brain Res 57:305–312

Montero VM (1991) A quantitative study of synaptic contacts on interneurons and relay cells of the cat lateral geniculate nucleus. Exp Brain Res 86:257–270

Moran J, Desimone R (1985) Selective attention gates visual processing in the extrastriate cortex. Science 23:782–784

Mulders WHAM, Robertson D (2000) Evidence for direct cortical innervation of medial olivocochlear neurons in rats. Hear Res 144:65–72

Murphy PC, Sillito AM (1987) Corticofugal feedback influences the generation of length tuning in the visual pathway. Nature 329:727–729

Murphy PC, Sillito AM (1996) Functional morphology of the feedback pathway from area 17 of the cat visual cortex to the lateral geniculate nucleus. J Neurosci 16:1180–1192

Murphy PC, Duckett SG, Sillito AM (1999) Feedback connections to the lateral geniculate nucleus and cortical response properties. Science 286:1552–1554

Murphy PC, Duckett SG, Sillito AM (2002) Comparison of the laminar distribution of input from areas 17 and 18 of the visual cortex to the lateral geniculate nucleus of the cat. J Neurosci 20:845–853

Murthy VN, Fetz EE (1992) Coherent 25- to 35-Hz oscillations in the sensorimotor cortex of awake behaving monkeys. Proc Natl Acad Sci U S A 89:5670–5674

Murthy VN, Fetz EE (1996) Synchronization of neurons during local field potential oscillations in sensorimotor cortex of awake monkeys. J Neurophysiol 76:3968–3982

Newberry NR, Simmonds MA (1984) The rat gracilis nucleus in vitro: III. Unitary spike potentials and their conditioned inhibition. Brain Res 303:59–65

Nicolelis MAL, Shuler M (2001) Thalamocortical and corticocortical interactions in the somatosensory system. Prog Brain Res 130:90–110

Nicolelis MAL, Baccala LA, Lin RCS, Chapin JK (1995) Sensorimotor encoding by synchronous neural ensemble activity at multiple levels of the somatosensory system. Science 268:1353–1358

Nuñez A, Amzica F (2004) Mecanismos de generación de las oscilaciones lentas del electroencefalograma durante el sueño. Rev Neurol 39:628–633

Nuñez A, Buño W (1999) In vitro electrophysiological properties of rat dorsal column nuclei neurons. Eur J Neurosci 11:1865–1876

Nunez A, Buño W (2001) Properties and plasticity of synaptic inputs to rat dorsal column neurones recorded in vitro. J Physiol (Lond) 535:483–495

Nuñez A, Panetsos F, Avendaño C (2000) Rhythmic neuronal interactions and synchronization in the rat dorsal column nuclei. Neuroscience 100:599–609

Oatman LC (1971) Role of visual attention on auditory evoked potentials in unanesthetized cats. Exp Neurol 32:341–356

Oatman LC, Anderson BW (1977) Effects of visual attention on tone burst evoked auditory potentials. Exp Neurol 57:200–211

Ohara PT, Lieberman AR (1981) Thalamic reticular nucleus: anatomical evidence that cortico-reticular axons establish monosynaptic contact with reticulo-geniculate projection cells. Brain Res 207:153–156

Ojima H (1994) Terminal morphology and distribution of corticothalamic fibers originating from layers 5 and 6 of cat primary auditory cortex. Cereb Cortex 4:646–663

Ojima H, Honda CN, Jones EG (1992) Characteristics of intracellularly injected infragranular pyramidal neurons in cat primary auditory cortex. Cereb Cortex 2:197–216

Orban GA (1994) Motion processing in monkey striate cortex. In: Peters A, Rockland KS (eds) Cerebral cortex, Vol 10. Plenum Press, New York, pp 431–441

Orban GA (1997) Visual processing in macaque area MT/V5 and its satellites MSTd and MSTv. In: Peters A, Rockland KS (eds) Cerebral cortex, Vol 13. Plenum Press, New York, pp 359–379

Orman SS, Humphrey GL (1981) Effects of changes in cortical arousal and of auditory cortex cooling on neuronal activity in the medial geniculate body. Exp Brain Res 42:475–482

Palmeri A, Bellomo M, Giuffrida R, Sapienza S (1998) Motor cortex modulation of exteroceptive information at bulbar and thalamic lemniscal relays in the cat. Neuroscience 88:135–150

Panetsos F, Nuñez A, Avendaño C (1997) Electrophysiological effects of temporary deafferentation on two characterized cell types in the nucleus gracilis of the rat. Eur J Neurosci 9:563–572

Panetsos F, Nuñez A, Avendaño C (1998) Sensory information processing in the dorsal column nuclei by neuronal oscillators. Neuroscience 84:635–639

Payne BR, Berman N (1984) Contralateral corticofugal projections to cat visual thalamus. Exp Brain Res 53:462–466

Pettit MJ, Schwark HD (1993) Receptive field reorganization in dorsal column nuclei during temporary denervation. Science 262:2054–2056

Popratiloff A, Valtschanoff JG, Rustioni A, Heinberg RJ (1996) Colocalization of GABA and glycine in the rat dorsal column nuclei. Brain Res 706:308–312

Popratiloff A, Rustioni A, Weinberg RJ (1997) Heterogeneity of AMPA receptors in the dorsal column nuclei of the rat. Brain Res 754:333–339

Priebe NJ, Ferster D (2005) Direction selectivity of excitation and inhibition in simple cells of the cat primary visual cortex. Neuron 45:133–145

Przybyszewski AW, Gaska JP, Foote W, Pollen DA (2000) Striate cortex increases contrast gain of macaque LGN neurons. Vis Neurosci 17:485–494

Puel JL, Bonfils P, Pujol R (1988) Selective attention modifies the active micromechanical properties of the cochlea. Brain Res 447:380–383

Raiguel S, Lagae L, Gulyas B, Orban GA (1989) Response latencies of visual cells in macaque areas V1, V2 and V5. Brain Res 493:155–159

Rapisarda C, Palmeri A, Sapienza S (1992) Cortical modulation of thalamo-cortical neurons relaying exteroceptive information: a microstimulation study in the guinea pig. Exp Brain Res 88:140–150

Rasmusson DD, Turnbull BG (1983) Immediate effects of digit amputation on S1 cortex in the racoon: unmasking of inhibitory fields. Brain Res 288:368–370

Reichova I, Sherman SM (2004) Somatosensory corticothalamic projections: distinguishing drivers from modulators. J Neurophysiol 92:2185–2197

Reynolds JH, Chelazzi L, Desimone R (1999) Competitive mechanisms subserve attention in macaque areas V2 and V4. J Neurosci 19:1736–1753

Ribary U, Ioannides AA, Singh KD, Hasson R, Bolton JPR, Lado F, Mogilner A, Llinas R (1991) Magnetic field tomography of coherent thalamocortical 40-Hz oscillations in humans. Proc Natl Acad Sci U S A 88:11037–11041

Rivadulla C, Martínez LM, Varela C, Cudeiro J (2002) Completing the corticofugal loop: A visual role for the corticogeniculate type 1 metabotropic glutamate receptor. J Neurosci 22:2956–2962

Rockland KS (1996) Two types of corticopulvinar terminations: round (type 2) and elongate (type 1). J Comp Neurol 368:57–87

Rockland KS, Knutson T (2000) Feedback connections from area MT of the squirrel monkey to areas V1 and V2. J Comp Neurol 425:345–368

Roelfsema PR, Engel AK, König P, Singer W (1997) Visuomotor integration is associated with zero-time lag synchronization among cortical areas. Nature 385:157–161

Rouiller EM, Duriff C (2004) The dual pattern of corticothalamic projection of the primary auditory cortex in macaque monkey. Neurosci Lett 358:49–52

Rouiller EM, Welker E (1991) Morphology of thalamic afferents from the auditory cortex of the rat; a Phaseolus vulgaris-leucoagglutinin PHA-L tracing study. Hear Res 56:179–190

Royce GJ (1982) Laminar origin of cortical neurons which project upon the caudate nucleus: a horseradish peroxidase investigation in the cat. J Comp Neurol 205:8–29

Rustioni A, Hayes NL (1981) Corticospinal tract collaterals to the dorsal column nuclei of cats. Exp Brain Res 43:237–245

Rustioni A, Cuénod M (1982) Selective retrograde transport of D-aspartate in spinal interneurons and cortical neurons of rats. Brain Res 236:143–155

Rustioni A, Sotelo C (1974) Synaptic organization of the nucleus gracilis of the cat. Experimental identification of dorsal root fibres and cortical afferents. J Comp Neurol 155:441–468

Rustioni A, Weinberg RJ (1989) The somatosensory system. In: Björklund A, Hökfelt T, Swanson LW (eds) Handbook of chemical neuroanatomy. Elsevier, Amsterdam, pp 219–221

Rustioni A, Schmechel DE, Spreafico R, Chema S, Cuénod M (1983) Excitatory and inhibitory amino acid putative neurotransmitters in the ventralis posterior complex: an autoradiographic and immunocytochemical study in rats and cats. In: Macchi G, Rustioni A, Spreafico R (eds) Somatosensory integration in the thalamus. Elsevier, Amsterdam, pp 365–383

Ryugo DK, Weinberger NM (1976) Corticofugal modulation of the medial geniculate body. Exp Neurol 51:377–391

Sakai M, Suga N (2001) Plasticity of the cochleotopic (frequency) map in specialized and nonspecialized auditory cortices. Proc Natl Acad Sci U S A 98:3507–3512

Sakai M, Suga N (2002) Centripetal and centrifugal reorganizations of frequency map of the auditory cortex in gerbils. Proc Natl Acad Sci U S A 99:7108–7112

Saldaña E, Merchán MA (2005) Intrinsic and commissural connections of the inferior colliculus. In: Winer JA, Schreiner CE (eds) The inferior colliculus. Springer, New York, pp 155–181

Saldaña E, Feliciano M, Mugnaini E (1996) Distribution of descending projections from primary auditory neocortex to inferior colliculus mimics the topography of intracollicular projections. J Comp Neurol 371:15–40

Sanchez-Vives MV, McCormick DA (1997) Functional properties of perigeniculate inhibition of dorsal lateral geniculate nucleus thalamocortical neurons in vitro. J Neurosci 17:8880–8893

Shatz CJ, Stryker MP (1988) Prenatal tetrodotoxin infusion blocks segregation of retinogeniculate afferents. Science 242:87–89

Scharfman HE, Lu SM, Guido W, Adams PR, Sherman SM (1990) N-methyl-D-aspartate (NMDA) receptors contribute to excitatory postsynaptic potentials of cat lateral geniculate neurons recorded in thalamic slices. Proc Natl Acad Sci U S A 87:4548–4552

Schiller PH, Tehovnik EJ (2001) Look and see: how the brain moves your eyes about. Prog Brain Res 134:127–142

Schiller PH, Tehovnik EJ (2003) Cortical inhibitory circuits in eye-movement generation. Eur J Neurosci 18:3127–3133

Schmielau F, Singer W (1977) The role of visual cortex for binocular interactions in the cat lateral geniculate nucleus. Brain Res 120:354–361

Schofield BR, Coomes DL (2004) Projections from the auditory cortex to the superior olivary complex in guinea pigs. Eur J Neurosci 19:2188–2200

Schofield BR, Coomes DL (2005) Projections from auditory cortex contact cells in the cochlear nucleus that project to the inferior colliculus. Hear Res 206:3–11

Schofield BR, Hallman LE, Lin CS (1987) Morphology of corticotectal cells in the primary visual cortex of hooded rats. J Comp Neurol 261:85–97

Schwartz ML, Dekker JJ, Goldman-Rakic PS (1991) Dual mode of corticothalamic synaptic termination in the mediodorsal nucleus of the rhesus monkey. J Comp Neurol 15:289–304

Schwartzkroin PA, Duijn H, Prince DA (1974) Effects of projected cortical epileptiform discharges on unit activity in the cat cuneate nucleus. Exp Neurol 43:106–123

Sherman SM, Guillery RW (1996) Functional organization of thalamocortical relays. J Neurophysiol 76:1367–1395

Sherman SM, Guillery RW (1998) On the actions that one nerve cell can have on another: distinguishing "drivers" from "modulators". Proc Natl Acad Sci U S A 95:7121–7126

Sherman SM, Guillery RW (2002) Thalamic relay functions and their role in corticocortical communication: generalizations from the visual system. Neuron 33:163–175

Sherman SM, Koch C (1986) The control of retinogeniculate transmission in the mammalian lateral geniculate nucleus. Exp Brain Res 63:1–20

Shipp S, Zeki S (1989) The organization of connections between areas V5 and V1 in macaque monkey visual cortex. Eur J Neurosci 1:309–332

Shosaku A, Sumitomo I (1983) Auditory neurons in the rat thalamic reticular nucleus. Exp Brain Res 49:432–442

Sillito AM, Jones HE (1997) Functional organization influencing neurotransmission in the lateral geniculate nucleus. In: Steriade M, Jones EG, McCormick DA (eds) Thalamus. Experimental and clinical aspects, Vol 2. Elsevier, Amsterdam, pp 1–52

Sillito AM, Jones HE (2002) Corticothalamic interactions in the transfer of visual information. Phil Trans R Soc Lond B 357:1739–1752

Sillito AM, Cudeiro J, Murphy PC (1993) Orientation sensitive elements in the corticofugal influence on centre surround interactions in the dorsal lateral geniculate nucleus. Exp Brain Res 93:6–16

Sillito AM, Jones HE, Gerstein GL, West DC (1994) Feature-linked synchronization of thalamic relay cell firing induced by feedback from the visual cortex. Nature 369:479–482

Singer W (1993) Synchronization of cortical activity and its putative role in information processing and learning. Ann Rev Physiol 55:349–374

Singer W (1999) Neuronal synchrony: a versatile code for the definition of relations? Neuron 24:49–65

Singer W, Gray CM (1995) Visual feature integration and the temporal correlation hypothesis. Ann Rev Neurosci 18:555–586

Singer W, Tretter F, Cynader M (1975) Organization of cat striate cortex: a correlation of receptive-field properties with afferent and efferent connections. J Neurophysiol 38:1080–1098

Slotnick SD, Moo LR, Kraut MA, Lesser RP, Hart JJr (2002) Interactions between thalamic and cortical rhythms during semantic memory recall in human. Proc Natl Acad Sci U S A 99:6440–6443

Smith AT, Singh KD, Greenlee MW (2000) Attentional suppression of activity in the human visual cortex. Neuroreport 7:271–277

Somogyi G, Hajdu F, Tombol T (1978) Ultrastructure of the anterior ventral and anterior medial nuclei of the cat thalamus. Exp Brain Res 31:417–431

Steinmetz PN, Roy A, Fitzgerald PJ, Hsiao SS, Johnson KO, Niebur E (2000) Attention modulates synchronized neuronal firing in primate somatosensory cortex. Nature 404:187–190

Stellwagen D, Shatz CJ (2002) An instructive role for retinal waves in the development of retinogeniculate connectivity. Neuron 33:357–367

Steriade M (1993) Cellular substrates of brain rhythms. In: Niedemyer E, Lopes da Silva F (eds) Electroencephalography. Williams and Wilkins, Baltimore, pp 27–62

Steriade M (2006) Grouping of brain rhythms in corticothalamic systems. Neuroscience 137:1087–1106

Steriade M, Timofeev I (2003) Neuronal plasticity in thalamocortical networks during sleep and waking oscillations Neuron 37:563–576

Steriade M, Gloor P, Llinas R, Lopes da Silva FH, Mesulam MM (1990) Basic mechanisms of cerebral rhythmic activities. Electroenceph Clin Neurophysiol 76:481–508

Steriade M, Contreras D, Curró-Dossi R, Nuñez A (1993a) The slow (<1 Hz) oscillation in reticular thalamic and thalamocortical neurons. Scenario of sleep rhythms generation in interacting thalamic and neocortical networks. J Neurosci 13:3284–3299

Steriade M, McCormick DA, Sejnowski TJ (1993b) Thalamocortical oscillations in the sleeping and aroused brain. Science 262:679–685

Steriade M, Amzica F, Contreras D (1996) Synchronization of fast (30–40 Hz) spontaneous cortical rhythms during brain activation. J Neurosci 16:392–417

Steriade M, Amzica F, Contreras D (1998) Dynamic properties of corticothalamic neurons and local cortical interneurons generating fast rhythmic (30–40 Hz) spike bursts. J Neurophysiol 79:483–490

Suga N (1984) The extent to which biosonar information is represented in the bat auditory cortex. In: Edelman GM, Gall WE, Cowan WM (eds) Dynamic aspects of neocortical function. John Wiley and Sons, New York, pp 315–373

Suga N, Ma X (2003) Multuparametric corticofugal modulation and plasticity in the auditory system. Nat Neurosci 4:783–794

Suga N, Zhang Y, Yan J (1997) Sharpening of frequency tuning by inhibition in the thalamic auditory nucleus of the mustached bat. J Neurophysiol 77:2098–2114

Suga H, Xiao Z, Ma X, Ji W (2002) Plasticity and corticofugal modulation for hearing in adult animals. Neuron 36:9–18

Sun XD, Jen PH, Sun DX, Zhang SF (1989) Corticofugal influences on the responses of bat inferior collicular neurons to sound stimulation. Brain Res 495:1–8

Sun X, Chen Q, Jen PH (1996) Corticofugal control of central auditory sensitivity in the big brown bat Eptesicus fuscus. Neurosci Lett 212:131–134

Sur M, Sherman SM (1982) Retinogeniculate terminations in cat: morphological differences between X and Y cell axons. Science 218:389–391

Swadlow HA, Weyand TG (1987) Corticogeniculate neurons, corticotectal neurons, and suspected interneurons in visual cortex of awake rabbits: receptive-field properties, axonal properties, and effects of EEG arousal. J Neurophysiol 57:977–1001

Tamamaki N, Uhlrich DJ, Sherman SM (1995) Morphology of physiologically identified retinal X and Y axons in the cat's thalamus and midbrain as revealed by intraaxonal injection of biocytin. J Comp Neurol 354:583–607

Temereanca S, Simons DJ (2004) Functional topography of corticothalamic feedback enhances thalamic spatial response tuning in the somatosensory whisker/barrel system. Neuron 41:639–651

Thomson AM (2000) Facilitation, augmentation and potentiation at central synapses. Trends Neurosci 23:305–312

Tolhurst DJ, Dean AF (1990) The effects of contrast on the linearity of spatial summation of simple cells in the cat's striate cortex. Exp Brain Res 79:582–588

Torterolo P, Pedemonte M, Velluti RA (1995) Intracellular in vivo recording of inferior colliculus auditory neurons from awake guinea-pigs. Arch. Ital. Biol 134:57–64

Towe AL, Jabbur SJ (1961) Cortical inhibition of neurons in dorsal column nuclei of cat. J Neurophysiol 24:488–498

Tsumoto T, Creutzfeldt OD, Legendy CR (1978) Functional organization of the corticofugal system from visual cortex to lateral geniculate nucleus in the cat. Exp Brain Res 32:345–364

Tsumoto T, Suda K (1980) Three groups of cortico-geniculate neurons and their distribution in binocular and monocular segments of cat striate cortex. J Comp Neurol 193:223–236

Turner JP, Salt TE (1998) Characterization of sensory and corticothalamic excitatory inputs to rat thalamocortical neurones in vitro. J Physiol. (Lond) 510:829–843

Turner JP, Salt TE (1999) Group III metabotropic glutamate receptors control corticothalamic synaptic transmission in the rat thalamus in vitro. J Physiol (Lond) 519:481–491

Usrey WM, Fitzpatrick D (1996) Specificity in the axonal connections of layer VI neurons in tree shrew striate cortex: evidence for distinct granular and supragranular systems. J Neurosci 16:1203–1218

Usrey WM, Reppas JB, Reid RC (1999) Specificity and strength of retinogeniculate connections. J Neurophysiol 82:3527–3540

Valverde F (1966) The pyramidal tract in rodents. A study of its relations with the posterior column nuclei, dorsolateral reticular formation of the medulla, and cervical spinal cord (Golgi and E.M. observations) Zeit Zellforsch 71:297–363

Van Horn SC, Erisir A, Sherman SM (2000) Relative distribution of synapses in the A-laminae of the lateral geniculate nucleus of the cat. J Comp Neurol 416:509–520

Van Horn SC, Sherman SM (2004) Differences in projection patterns between large and small corticothalamic terminals. J Comp Neurol 475:406–415

Vanduffel W, Tootell RB, Orban GA (2000) Attention-dependent suppression of metabolic activity in the early stages of the macaque visual system. Cereb Cortex 10:109–126

Veinante P, Lavallée P, Deschênes M (2000) Corticothalamic projections from layer 5 of the vibrissal barrel cortex in the rat. J Comp Neurol 424:197–204

Vetter DE, Saldana E, Mugnaini E (1993) Input from the inferior colliculus to medial olivocochlear neurons in the rat: a double label study with PHA-L and cholera toxin. Hear Res 70:173–186

Vidnyanszky Z, Gorcs TJ, Negyessy L, Borostyankio Z, Knopfel T, Hamori J (1996) Immunocytochemical visualization of the mGluR1a metabotropic glutamate receptor at synapses of corticothalamic terminals originating from area 17 of the rat. Eur J Neurosci 8:1061–1071

Vidyasagar TR (1998) Gating of neuronal responses in macaque primary visual cortex by an attentional spotlight. Neuroreport 9:1947–1952

Villa AE, Rouiller EM, Simm GM, Zurita P, de Ribaupierre Y, de Ribaupierre F (1991) Corticofugal modulation of the information processing in the auditory thalamus of the cat. Exp Brain Res 86:506–517

Volgushev M, Voronin LL, Chistiakova M, Singer W (1997) Relations between long-term synaptic modifications and paired-pulse interactions in the rat neocortex. Eur J Neurosci 9:1656–1665

Walberg F (1966) The fine structure of the cuneate nucleus in normal cats and following interruption of afferent fibres. An electron microscopical study with particular reference to findings made in Glees and Nauta sections. Exp Brain Res 2:107–128

Waleszczyk WJ, Wang C, Burke W, Dreher B (1999) Velocity response profiles of collicular neurons: parallel and convergent visual information channels. Neuroscience 93:1063–1076

Waleszczyk WJ, Bekisz M, Wróbel A (2005) Cortical modulation of neuronal activity in the cat's lateral geniculate and perigeniculate nuclei. Exp Neurol 196:54–72

Wang C, Waleszczyk WJ, Benedek G, Burke W, Dreher B (2001) Convergence of Y and non-Y channels onto single neurons in the superior colliculi of the cat. Neuroreport 12:2927–2933

Watanabe M, Mishina M, Inoue Y (1994) Distinct distributions of five NMDA receptor channel subunit mRNAs in the brainstem. J Comp Neurol 343:520–531

Watanabe T, Yanagisawa K, Kanzaki J, Katsuki Y (1966) Cortical efferent flow influencing unit responses of medial geniculate body to sound stimulation. Exp Brain Res 2:302–317

Webb BS, Tinsley CJ, Barraclough NE, Easton A, Parker A, Derrington AM (2002) Feedback from V1 and inhibition from beyond the classical receptive field modulates the responses of neurons in the primate lateral geniculate nucleus. Vis Neurosci 19:583–592

Weedman DL, Ryugo DK (1996a) Projections from auditory cortex to the cochlear nucleus in rats: synapses on granule cell dendrites. J Comp Neurol 371:311–324

Weedman DL, Ryugo DK (1996b) Pyramidal cells in primary auditory cortex project to cochlear nucleus in rat. Brain Res 706:97–102

Weimann JM, Zhang YA, Levin ME, Devine WP, Brulet P, McConnell SK (1999) Cortical neurons require Otx1 for the refinement of exuberant axonal projections to subcortical targets. Neuron 24:819–831

Weinberger NM (1998) Physiological memory in primary auditory cortex: characteristics and mechanisms. Neurobiol Learn Mem 70:226–251

Weisberg JA, Rustioni A (1976) Cortical cells projecting to the dorsal column nuclei of cats. An anatomical study with the horseradish peroxidase technique. J Comp Neurol 168:425–438

Weisberg JA, Rustioni A (1979) Differential projections of cortical sensorimotor areas upon the dorsal column nuclei of cats. J Comp Neurol 184:401–422

White EL, Hersch SM (1982) A quantitative study of thalamocortical and other synapses involving the apical dendrites of corticothalamic projection cells in mouse SmI cortex. J Neurocytol 11:137–157

Williamson AM, Ohara PT, Ralston HJ 3rd (1993) Electron microscopic evidence that cortical terminals make direct contact onto cells of the thalamic reticular nucleus in the monkey. Brain Res 631:175–179

Wilson JR, Friendlander MJ, Sherman SM (1984) Fine structural morphology of identified X- and Y-cells in the cat's lateral geniculate nucleus. Proc R Soc Lond B 221:411–436

Winer JA, Larue DT (1987) Patterns of reciprocity in auditory thalamocortical and corticothalamic connections: study with horseradish peroxidase and autoradiographic methods in the rat medial geniculate body. J Comp. Neurol 257:282–315

Winer JA, Prieto JJ (2001) Layer V in cat primary auditory cortex (AI): cellular architecture and identification of projection neurons. J Comp Neurol 434:379–412

Winer JA, Larue DT, Diehl JJ, Hefti BJ (1998) Auditory cortical projections to the cat inferior colliculus. J Comp Neurol 400:147–174

Winer JA, Diehl JJ, Larue DT (2001) Projections of auditory cortex to the medial geniculate body of the cat. J Comp Neurol 430:27–55

Wise SP, Jones EG (1978) Developmental studies of thalamocortical and commissural connections in the rat somatic sensory cortex. J Comp Neurol 178:187–208

Xiao Z, Suga N (2002) Modulation of cochlear hair cells by the auditory cortex in the mustached bat. Nat Neurosci 5:57–63

Xiao Z, Suga N (2005) Asymmetry in corticofugal modulation of frequency-tuning in mustached bat auditory system. Proc Natl Acad Sci U S A 102:19162–19167

Xiong Y, Yu YQ, Chan YS, He J (2004) Effects of cortical stimulation on auditory-responsive thalamic neurons in anesthetized guinea pigs. J Physiol (Lond) 560:207–217

Yan J, Ehret G (2001) Corticofugal reorganization of the midbrain tonotopic map in mice. Neuroreport 12:3313–3316

Yan J, Ehret G (2002) Corticofugal modulation of midbrain sound processing in the house mouse. Eur J Neurosci 16:119–128

Yan J, Suga N (1996) Corticofugal modulation of time-domain processing of biosonar information in bats. Science 273:1100–1103

Yan W, Suga N (1998) Corticofugal modulation of the midbrain frequency map in the bat auditory system. Nat Neurosci 1:54–58

Yeterian EH, Pandya DN (1994) Laminar origin of striatal and thalamic projections of the prefrontal cortex in rhesus monkeys. Exp Brain Res 99:83–98

Yu YQ, Xiong Y, Chan YS, He J (2004) In vivo intracellular responses of the medial geniculate neurones to acoustic stimuli in anaesthetized guinea pigs. J Physiol (Lond) 560:191–205

Yuan B, Morrow TJ, Casey KL (1985) Responsiveness of ventrobasal thalamic neurons after suppression of S1 cortex in the anesthetized rat. J Neurosci 5:2971–2978

Yuan B, Morrow TJ, Casey KL (1986) Corticofugal influences of S1 cortex on ventrobasal thalamic neurons in the awake rat. J Neurosci 6:3611–3617

Zahs KR, Stryker MP (1988) Segregation of ON and OFF afferents to ferret visual cortex. J Neurophysiol 59:1410–1429

Zhang Y, Suga N (2000) Modulation of responses and frequency tuning of thalamic and collicular neurons by cortical activation in mustached bat. J Neurophysiol 84:325–333

Zhang Y, Suga N (2005) Corticofugal feedback for collicular plasticity evoked by electric stimulation of the inferior colliculus. J Neurophysiol 94:2676–2682

Zhang Y, Suga N, Yan J (1997) Corticofugal modulation of frequency processing in bat auditory system. Nature 387:900–903

Zhou X, Jen PH (2000) Brief and short-term corticofugal modulation of subcortical auditory responses in the big brown bat, Eptesicus fuscus. J Neurophysiol 84:3083–3087

Zucker RS (1989) Short-term synaptic plasticity. Ann Rev Neurosci 12:13–31

Subject Index

GPSR Compliance

The European Union's (EU) General Product Safety Regulation (GPSR) is a set of rules that requires consumer products to be safe and our obligations to ensure this.

If you have any concerns about our products, you can contact us on ProductSafety@springernature.com

In case Publisher is established outside the EU, the EU authorized representative is:

Springer Nature Customer Service Center GmbH
Europaplatz 3
69115 Heidelberg, Germany

The manufacturer's authorised representative in the EU is Springer
Nature Customer Service Centre GmbH, Europaplatz 3, 69115 Heidelberg,
Germany. If you have any concerns regarding our products, please
contact ProductSafety@springernature.com

Printed and bound by CPI Group (UK) Ltd, Croydon, CR0 4YY

07/01/2026

02030933-0006